• 国家自然科学基金重点项目（51138004）
• 华南理工大学亚热带建筑科学国家重点实验室开放课题项目（2016ZB17）
• 《广州大典》与广州历史文化研究资助项目（2015GZY25）
• 华南理工大学出版基金资助项目

自然化建筑：形态与肌理

N-ArchitecTURE: Morphology & Spatial Pattern

凌晓红 著

·广州·

前　言

人类社会自有了建筑，建筑与自然的关系就成了探讨的永恒主题。建筑最初产生的目的是人类用以抵抗自然界带来的侵害。然而，在与自然长期互动的过程中，人们发现，自然界不仅为人类的建成环境提供了一个背景与舞台，而且在技术和材料方面也是人类完成建造最大的启发者与供应者。因此，从某种意义上说，建筑是人类与自然互动关系的一种表现。有鉴于此，建筑设计应回归自然的本质。

传统的设计观念倾向于把城市和建筑看成对立于自然界的一种人造环境；然而，建筑是自然的一部分，乃至于就是自然景观本身，却是一直以来伴随建筑学发展的一种重要主张。尤其在 20 世纪 80 年代，当可持续发展的观念开始盛行，建筑与自然的关系更是超越原有的理解和定义，不断地更新与演变，为建筑设计带来无限的可能性。过去，我们主要在城市尺度讨论自然化设计的问题；然而，随着各学科交叉的盛行，自然化设计就渗透至各个尺度的建成环境。如何有机处理建筑与自然景观的关系常常成为设计的主题，也成为解决特定建筑命题的重要手段。

本书通过提出“自然化建筑”(N-architecture) 的概念，对与自然化相关的建筑学理论进行总结与梳理，深入探讨自然与景观、景观与建筑关系的历史演变，并以先锋性的设计竞赛作品为案例，探讨多元的设计观念，分析它们可能存在的形态、空间肌理，以及付诸实施的具体策略与手段。

Preface

Since the emergence of architecture, its relationship with nature has been a topic pioneered in society. Originally, architecture was an attempt to protect people against the infringements of nature. However, in the course of long-term interaction with nature, people realized that nature has not only provided their built environment with a place or background, but also functioned as a source of inspiration and a generous supplier for architectural construction in terms of technology and materials. In this sense, architecture should be a reflection of the interaction between humans and nature, and thus architectural design should be initiated from nature.

Traditionally, cities and buildings were treated as the opposite to nature; however, the proposition that architecture is an integral part of nature and even the landscape itself has accompanied the development of architectural science over time. Notably in the 1980s, when sustainable development began to prevail, the understanding of the relationship between architecture and nature went beyond its conventional scope, constantly transforming and evolving, subsequently generating countless possibilities in design practice. In the past, the naturalized design issue was mainly discussed on the urban scale; yet, under the influence of interaction among different disciplines, naturalized architecture (defined as 'N-architecture' in this book) has become an important means to solve particular architectural issues.

Through initiating the concept of N-architecture, this book first reviews related theories in the literature, and then analyzes the historical evolution of the relationship between nature and landscape, and between landscape and architecture. After that, this book explores diverse design ideas through the studies of a number of pioneering architectural works, subsequently identifying potential relational modes, design strategies, and spatial patterns that they can produce.

目　录

Contents

Staring......
Sculpture Studio
Caricature studio
Funny games !
Arts on sale !

绘制：杨森博、林丹薇

Welcome to our first exhibition !

I wanna buy it !

I enjoy the exhibition under the trees !

Painting Studio

What a wonderful fashion mart !

Porcelain Studio

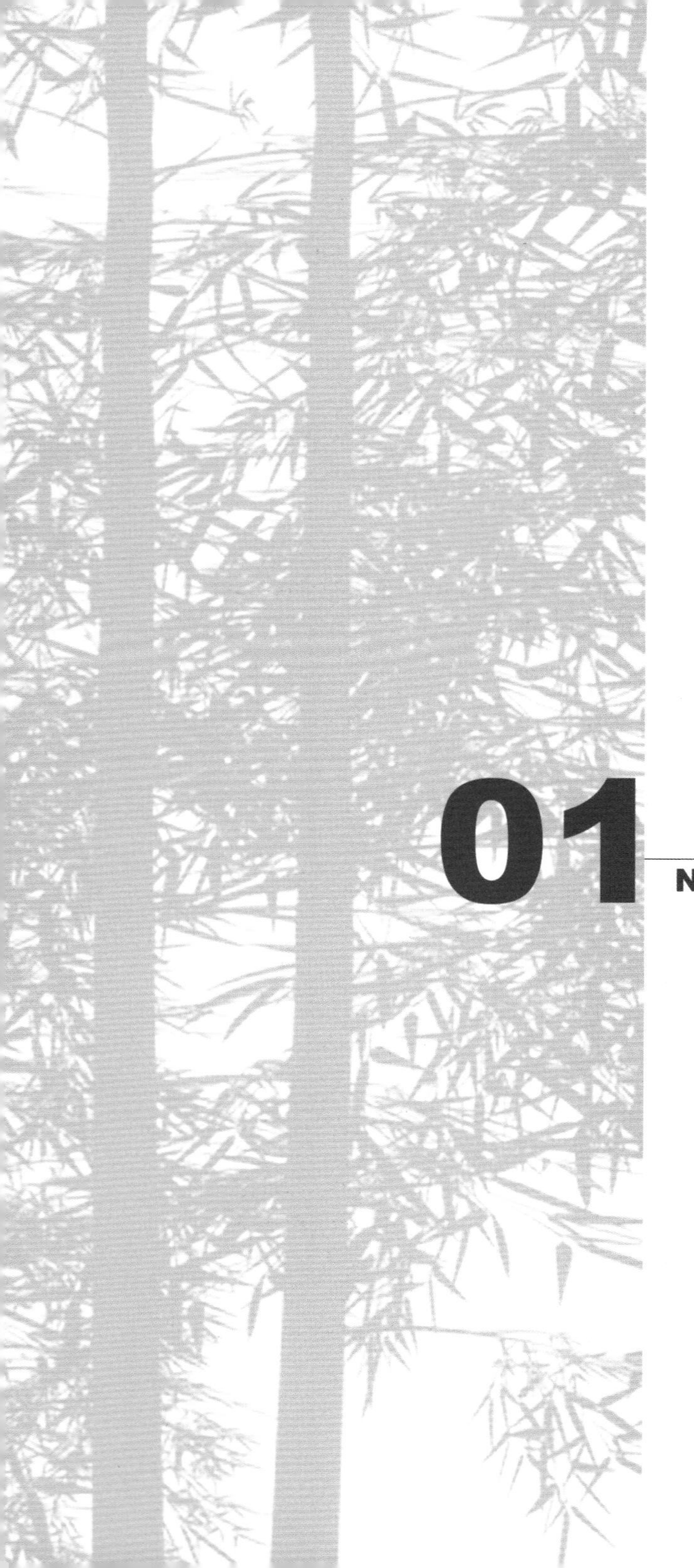

01 自然化建筑及其理论基础

N-architecture and its Theoretical Foundation

1.1 自然化建筑设计主张

1.1.1 概论

“人类是一种有思维能力和情感的独特生物，他们可以感知和表达情感，因而成为自然界范式的制定者。然而，要完成这使命，人类的设计行为必须基于自然。”

—— 麦克哈格

麦克哈格指出人的自然属性，从而为人的创造性活动回归于自然埋下了伏笔。建筑学科发展至今，设计的目的已经超越仅仅创建一栋完美的建筑物这么一个简单的命题。建筑不再是以自我为导向的物体，它与城市及周围环境的关系受到越来越多的关注，并促使人们热切地探讨：自然如何可以有效地成为一种语言，融合在设计的思考当中。当代大量的设计实践已经表明，在跨学科的背景下这种融合关系会产生无数的可能性。

近一个世纪以来，基于建筑自然化这种源动力，建筑学领域已经出现大量新的思潮和设计理念；然而，系统阐述这些观点的文献并不多见。因此，本书在提出“自然化建筑”概念的基础上，从理论层面对自然化建筑的形成与发展过程进行简述，并在实例分析的基础上归纳总结出一系列的空间形态与肌理，以期为设计实践提供参考与借鉴。

1.1 N-architecture: Design with Nature

1.1.1 Introduction

McHarg said, 'Man is that unique conscious creature who can perceive and express. He must become the standard of the biosphere. To do this, he must design with nature.'

McHarg pointed out that the inherent characteristic of humans is that their creative activities can return to nature. Nowadays, architectural design has evolved beyond the mere purpose of creating a perfect single building, whatever it may be. Architecture is no longer deemed a self-orientated entity without paying attention to its relationship with the natural environment. In this sense, designers need to explore how nature can be, as a design language, effectively involved in the design process. It is worth noting that this kind of integration could produce numerous design possibilities in light of the interdisciplinary initiatives.

Driven by the naturalized forces over the past century, there have emerged a large number of innovative design ideas. However, little of the literature expounds these notions systematically. Therefore, this book is an attempt to redefine the concept of 'N-architecture', by reviewing its formation and development at the theoretical level, subsequently identifying a series of relational schema in order to provide contemporary design practice with valuable guidance and inspirations.

1.1.2 本书的结构

本书分为三个章节。第一章先对自然化建筑的理论发展进行简要回顾，从而揭示自然化观念一直伴随着建筑学的发展，反映人类对所处环境的思考从未间断，并在此过程中交织着敬畏、妥协、赞美、互动等复杂情感。

第二章主要从形态学的角度分析建筑与自然景观可能产生的关系模式，并通过对相关设计实例的分析，展示不同关系模式下可能产生的建筑现象。

第三章则从空间肌理的角度，展示各种自然化建筑理念在转化为物质空间过程中，通过一定的手段和策略，构建出建筑、人、自然之间复杂的空间关系。

1.1.2 Structure of the Book

This book comprises three sections. The first chapter gives a brief review of the theoretical development of N-architecture, revealing a fact that the concept of naturalization has been accompanied by the progress and development of architecture over a long period of time. During this process, people have never stopped thinking about their relationship with the natural environment, as well as mingling with various kinds of emotions, including fear, surrender, praise and interaction.

The second part of this book intends to explore the possible relationship between architecture and natural landscape mainly from the morphological point of view. Through the analysis of a number of design examples, different spatial relational patterns are identified and illustrated.

Contrastingly, the third part of this book reveals the strategies and methods to achieve N-architecture mainly based on various spatial patterns that result in different relationships between architecture, humans and nature.

1.2 建筑景观化与景观建筑化

1.2.1 自然与景观

回顾人类社会发展的过程，城市规模不断扩张，人口急剧膨胀。人类的建成环境一直在侵蚀着自然环境，从而使什么是自然的问题开始变得模糊。语义上，自然本是美的代名词。在定义什么是美的问题时，著名哲学家康德就曾指出，当人类所有的心智都在与自然发生互动时就产生了美。然而，人们常常认为一个城市的天际线很美，它却不是自然的。康德最后解释说，广义的自然应包括人类所创造的一切（包括城市和建筑），这属于第二自然，因为人类是大自然的一部分，所以拥有自然界的属性。这就意味着，人类一切活动包括设计活动，都离不开自然。

本书提出的“自然化建筑”可以说是在创造一种景观化建筑或是一种建筑化景观；然而，它所讨论的内容并不等同于当代景观建筑学所涉及的范畴。因此，在进一步探讨景观与建筑的问题时，我们有必要先厘清景观与自然的关系。景观，即是自然却又不等同于自然。

1.2 Architecturalized Landscape and Landscaped Architecture

1.2.1 Nature vs. Landscape

Reviewing the history of human society, it is easy to find that with the rapid population growth and continuous expansion of cities, artificial environment has been swallowing up the natural setting constantly. Consequently, nature is not like nature, and artifacts are not like artifacts. The question as to what nature thus has become is vague. Conceptually, nature is always related to the sense of beauty. When defining what beauty is, Kant claimed that beauty occurs when all our mental faculties are in play with nature. Ironically, we consider a city's skyline as a kind of beauty, yet it is not natural. Finally, Kant explained that nature comprises what humans have created (including cities and architecture)—the second nature—because humans are part of nature, as well as the purpose of nature. In this sense, human activities can never be separated from nature.

Generally, N-architecture defined in this book can be regarded as the creation of landscaped architecture or architecturalized landscape. However, it is not equivalent to the scope and content that contemporary landscape architecture covers. To further explore the issue concerning landscape and architecture, there is a need to understand what landscape is and how it relates to nature. Landscape belongs to nature, yet it is not the same as nature.

1.2.2 景观是人与自然关系的人文表达

在西方，与“景观”类似的词最早出现于公元前的旧约《圣经》，希伯来文为“Noff”，词源上与“Yafe”即“美丽”有关。文中，“Noff”用以描述耶路撒冷壮丽的景色（图1-1）；由此可见，景观最早的含义是城市景观。而当代采用的“景观”（Landscape）这一词汇始于16世纪之交，是从罗马和日耳曼文字衍生出来的，当时主要用以描述以自然风光为主题的绘画作品。在环境学科领域，关于什么是景观的问题在历史上已经有不少的讨论，甚至出现相互对立的观点，至今尚未有一个精确的定义被广泛接受。但随着时间的推移，景观的内涵也开始发生变化。Coones曾指出，鉴于问题的复杂特性和混淆冲突的一些诠释，景观这一术语长久以来就包含各种内容，并融合丰富复杂的概念、美学观及学术追求。Coones并没有因为景观是一个人类创造的环境而把它从自然界中分离出来；相反，他认为景观是人类与自然环境关系的一种表达。某种程度上，景观反映了特定社会环境中文化体系的运作模式。景观在本质上是复杂的，是物质环境相互调和的产物，是不见痕迹的自然界演化过程与人类社会文化活动相互渗透的结果。而杰克森也认为：“景观是人们有意识地推进或阻碍自然过程的空间。”他认为景观不仅仅是被观赏的对象，而是存在于人类生活当中，是一种社会生活空间，是人与自然的有机整体。

1.2.2 Landscape as Humanity's Interpretation of Nature

In the West, the original word referring to 'Landscape' can be traced back to the Old Testament, as 'Noff' in Hebrew, which etymologically was related to 'Yafe' meaning 'Beauty'. In the Old Testament, 'Noff' was used to describe the spectacular views of Jerusalem, implying that the earliest meaning of landscape was related to the scenery of cities.

By contrast, the modern concept of 'landscape' was derived from both Romanticism and Germanism around the turn of the 16th century. Initially, it was used to denote the paintings whose primary subject matters were about natural scenery. All along, as for the definition of landscape, it has been answered in many different and even mutually exclusive ways. Currently, there is no certain answer or definition that has been widely accepted. Incidentally, with the passing of time, the meaning of landscape has plainly changed. Being confused by the conflicting interpretations and their failure to reflect the complexity of the subject, Coones pointed out that the definition of landscape has for long been variously employed to embrace a wide range of conceptions, definitions, aesthetic visions and academic pursuits. Coones did not regard landscape as the creation of society being divorced from nature, but as an expression of the relationship between humans and the natural world, as well as the operation of social life within particular milieu. He further stated that landscape is in truth nothing less than the complex, interrelated and unified material product of a geographical environment and seamless totality, in which the immemorial process of nature and the much recent act of mankind interpenetrate. Another author Jackson also said that landscape is the space where people intentionally impel or impede the natural process. In his opinion, landscape is not just an object to be observed, but exists in humans' lives. It is a social living space, and an organic whole involving both humans and nature.

图1-1 旧约《圣经》用“Noff”描述耶路撒冷壮丽的景色

1.2.3 建筑与景观的学科融合

景观作为一门独立学科出现的历史并不久远。19 世纪 50 年代，被称为美国景观建筑学之父的奥姆斯特德和英国建筑师沃克斯设计的纽约中央公园（建于 1938 年）的建造（图 1-2），标志景观设计开始进入普通公民的生活；并以此为始点，景观建筑学科在美国正式成立，标志着人类对建筑与自然的关系更加重视。人们试图在它们之间找到平衡点，以此来阐述两者错综复杂、既分又合的关系。同时，随着美国社会的现代化和民主化的进程，景观设计也逐渐从传统的、主要服务于少数人群的操作转变成涉及人类社会生活各个尺度的学科。

1.2.3 Disciplinary Integration between Architecture and Landscape

The history of the subject of Landscape Architecture is not long. In the 1850s, the father of American landscape architecture Frederick Law Olmsted and British Architect Calvert Vaux designed the Central Park of New York, symbolizing that landscape design began to enter into the lives of ordinary citizens. Also, this was the starting point when the subject of landscape architecture was formally established in the United States, implying that the relationship between nature and architecture has been paid increasing attention, and that people intended to interpret their complicated and interconnected relationships in various ways. Meanwhile,

图1-2 纽约中央公园

19世纪后期，现代主义建筑开始萌芽，并于20世纪初期进入繁盛时期。现代主义对景观建筑学的贡献不仅体现在新材料、新技术的运用上，而且认为功能化是景观设计的起点，从而使景观设计摆脱某种风景式的先验主义，重视与场地与建筑的结合及适应性设计。随着科学的发展和技术的进步，学科发展也日趋多元及出现学科交叉，景观建筑化或建筑景观化进一步加剧，从而使两个学科的融合达到前所未有的地步。在此大背景下，借助于生态、文化、美学、社会等多种学科的支撑，建筑景观化的实践也逐渐成型，并在建筑学领域形成了一个具有一定学术发展潜力的分支（图1-3~图1-5）。

20 世纪 70 年代，全球的生态和环境问题日益凸显，宾夕法尼亚大学景观建筑学教授麦克哈格因而提出将景观作为一个包括地质、地形、水文、土地利用、植物、气候等决定性要素及相互关联的整体来看待的观点，再一次扩展景观设计的范畴与内容，使它与其他环境要素的结合更加紧密。当大尺度景观实践逐渐转向理性科学的生态设计时，小尺度的景观设计由于受 60 年代以来环境艺术的影响及后现代思潮的熏陶，其内涵也发生了变化。景观设计因而突破传统自然主义的限制，而把注意力放在人造空间的集成上，并与城市公共生活紧密结合。其结果是，景观设计不再局限于纯自然要素（如植物、地形）的组合与操作，而更趋向于多元化，景观化建筑和建筑化景观在此情形下也被纳入景观建筑学讨论的范畴（如地景化建筑），从而使自然化建筑在建筑学领域逐渐成为一个特殊的分支。

accompanied by the modernization and democratization of the American society, the practice of landscape design was gradually transformed from primarily serving the minorities into a discipline involved in various dimensions of social life.

In the late nineteenth century, modernism began to sprout, and prospered in the early twentieth century. The contribution of modernism to landscape architecture lay not only in the availability of new materials and technologies, but also in the influence of the ideology that function was the starting point of landscape design. In light of the concept, landscape design got rid of a certain type of transcendentalism and focused more on its interconnection with architectural spaces. Additionally, with the advancement of science and technology, landscape architecture increasingly featured on multivariate and interdisciplinary, and the development of landscaped architecture and architecturalized landscape was further strengthened, resulting in the integration of two disciplines to an unprecedented extent. As a result, being supported by ecological, cultural, aesthetic, social and other disciplines, N-architecture has become one of the most valuable research cores in the field of architecture.

Since the 1970s, global warming and other environmental problems have become increasingly prominent. Ian L. McHarg, the founder of Department of Landscape Architecture at the University of Pennsylvania, thus promoted landscape architecture as

图1-3　传统的景观设计

图1-4　现代景观设计

图1-5　地景建筑：济州山酒店度假村

an interrelated system, involving geology, topography, hydrology, land use, vegetation, climate and other critical elements as a whole, further expanding the scope and content of landscape design, in addition to making it more closely related to other environmental factors.

While the large-scale landscape practice has gradually turned into an ecologically scientific approach, the meaning of small-scale landscape design has also been changed due to the influence of environmental arts and post-modern ideas prevailing in the 1960s. By breaking the traditional limitations of naturalism, landscape practice tended to focus more on the organization of architectural spaces and its tight integration with city public life. As a result, landscape design was no longer constrained to the arrangements of pure natural elements, including plants and topography, but became more diversified. In this case, landscape-oriented architecture and architecturalized landscape emerged and increasingly aroused public debates and discussion in architectural science, such as landform buildings. Gradually, N-architecture has turned into a particular typology in the field of architecture.

1.3 自然化建筑理论基础的演变与发展

其实，自然化建筑并不是在当代才出现的。回顾建筑学发展的历史，不难发现，建筑学理论的发展就一直包含自然化和景观化的主题，反映人类对所处环境的思考从未间断过，并在此过程中交织着敬畏、妥协、赞美、互动等复杂情感。

1.3.1 图形 – 背景原理

图形 – 背景原理是历史上较早探讨城市建成环境与自然环境相互关系的一种方法。根据格式塔理论，人类视觉感知可分为两个部分：一个是“图形”（视觉主要关注的对象），一个是“底”（除图以外被感知的背景部分）；图与底各自拥有自己的视觉特点，并在人类视觉感知的过程中起不同作用。当图底原理应用于建筑和城市设计领域时，常用以分析建筑实体空间相对于周围环境的形态及比例关系。

较好展示图底原理运用的例子是 Nolli 于 1748 年绘制的罗马地图（图 1–6），Nolli 首次用城市公共空间结构和市民的可达性来重新诠释罗马城市。这张地图清晰地向人们展示了城市形态是由建筑实体及实体之间一个连续的空间系统组成的。由于建筑实体覆盖的部分远比虚的外部空间紧凑密集，从而起到界定外部空间形态的作用。

1.3 Theoretical Evolution and Development of N-architecture

N-architecture did not emerge merely in the contemporary era. Reviewing the history of architectural development, it is found that architectural theories have constantly contained relevant themes, demonstrating that humans have never stopped thinking about their relationship with the natural environment, and in this process being mingled with complicated emotions, including fear, surrender, praise and interaction.

1.3.1 Figure-ground Principle

Historically, figure-ground was an earlier principle used to explore the relationship between architectural environments and external spaces. As derived from Gestalt theory, visual perception can be divided into two parts—a figure (the object of attention) and a ground (the remainder of the perception). Phenomenologically, these two sub-fields are different, with each one having its own characteristics and playing a different role in humans' visual perception. When applied to the field of architecture or urban design, the figure-ground principle has been acknowledged as an effective tool to analyze the shape and proportion of buildings as a solid-mass (figure) to the surrounding environment (ground).

The best illustration of the figure-ground principle in history was the map of Rome drawn by Giambattista

在这类城市环境中，外部空间体现出与建筑很深的集成连贯性，并在各具特色的街区及建筑群之间形成一个网格结构。也就是说，虚的外部空间俨如从建筑实体中雕刻出来一样，构成一个容纳各种社会活动，并具有流动性的空间体系。从构图上而言，虚的外部空间在视觉感知上反而成了积极的要素，比界定它的建筑实体更具有图案感，更易为人的视觉所感知。然而，图形之所以成为一个积极因素并引起视觉的注意，是因为背景的衬托与对比作用。图与底，并不是对立体，而是相互依存的整体。Nolli的罗马地图第一次以开放空间而非建筑为主体，重新思考城市空间的构成关系，说明在城市发展的过程中，人们逐步认识到建筑并非独立于环境的个体；相反，更能引起颂赞的是建筑或建筑组群之间的空间，因为它们是社会活动的载体和城市活力的源泉。

众所周知，欧洲城市的传统街区常以充满活力的街道空间及各类广场著称，而其城市肌理也往往清晰细腻，体现公共与私有领地的均匀相间、内外空间互为界定的形态特点，反映人们在造城过程中，已拥有利用外部空间构建生活场所的智慧与策略。相反，一些粗糙的城市肌理，图底关系往往趋向于破碎、片段化和孤立，这在现代主义的规划城市中并不鲜见（图1–7）。

Nolli in 1748, which was an earlier attempt to reinterpret the city of Rome by illustrating the accessibility of citizens and the structure of public spaces. Through revealing the city as a clearly defined system of solids and voids, the map illustrated that the building coverage was much denser than that of exterior spaces, ultimately playing a role in defining the shape of the public realms. In other words, buildings created voids, and the voids seemed to be carved out of buildings as a continuous flow linking exterior and interior spaces and activities. As a result, the outdoor spaces appeared as positive, and were more figural than the solids defining them. However, it is worth noting that figures, the positive part that attracts our attention, cannot exist without a contrasting background. Figure and background, therefore, are more than opposing elements. Rather, they form a unity of opposites. By some means, the map of Rome has indicated that during the process of urban development, architecture has not been deemed an entity independent of the surrounding environment; on the contrary, it is precisely the spaces between buildings or groups of buildings that are worthy of appreciating because they are the source of urban vitality.

As we all know, traditional European towns are famous for their vibrant streets and squares, and their urban fabric is always distinct and delicate, reflecting a clear relationship between the public and private realms. Additionally, the morphological characteristics of inner and external spaces are mutually defined. By contrast, with respect to the rough urban fabric, the figure-ground relationship tends to be fragmented and isolated, which is fairly commonly found in modern cities.

1.3.2 现代主义建筑与景观理念

19 世纪后期，现代主义建筑思潮开始萌芽和发展。1851 年建造的伦敦水晶宫和 1889 年的巴黎埃菲尔铁塔是现代主义萌芽的标志。1926 年，格雷皮乌斯在德国设计以及建造的包豪斯校舍标志现代主义建筑进入成熟时期(图 1-8)。在这个历史阶段，建筑实践普遍强调功能主义和技术理性，从而使建筑学科发展进入一个崭新时期。

1.3.2 Modernist Architecture and Landscape Concepts

In the late nineteenth century, the ideology of modernism began to develop. Crystal Palace built in London in 1851 and the Eiffel Tower in Paris completed in 1889 are two examples of modernist architecture. In 1926, Walter Gropius established the Bauhaus School in Germany, representing that modernist architecture had entered its mature period. During this period, architectural practice empha-

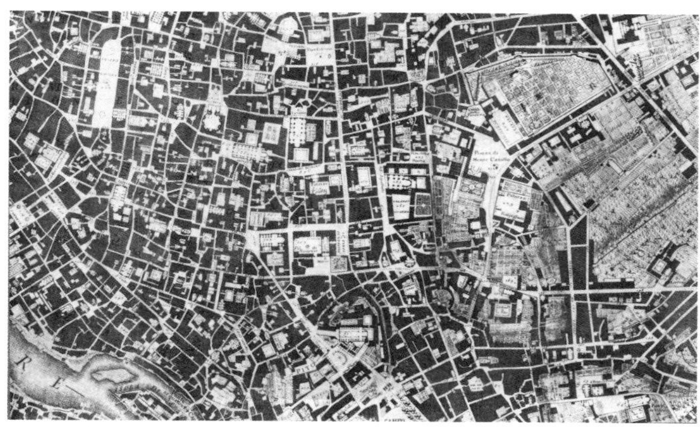

图1-6 Nolli于1748年绘制的罗马地图

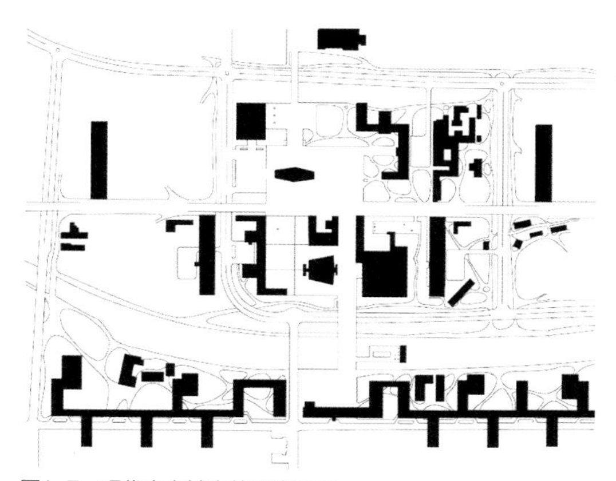

图1-7 现代主义城市的图底关系

然而，到了20世纪40至60年代，现代主义建筑和国际主义风格基本垄断了建筑设计实践，地方性、民族性在逐渐消退，城市面貌逐渐走向呆板、单调。

sized functionalism and technical rationality, leading the development of the discipline into a novel era. From the 1940s to 1960s, modernist architecture and its derived international style dominated the architectural practice of the world, with localism and nationalism gradually subsiding. Increasingly, modern cities became dull and monotonous.

图1-8　伦敦水晶宫（左上），巴黎埃菲尔铁塔（右）和包豪斯校舍（左下）

在现代主义规划思想引导下的城市发展，建筑往往呈现独立于景观环境的一种物质形态，而不再像传统街区那样，是街道、广场及其他开放空间组成的空间系统的重要组成部分，城市中也因而出现许多缺乏界定的“反空间”，即建筑之间形成尺度超然和无形的开放空间，建筑与环境之间似乎丧失了一种张力和内在联系性。诚然，现代建筑大师柯布西耶曾倡导，“人们通过阳光、空气、花草、树木，从而在自然中得到精神上的愉悦和满足感”。然而，这种主张某种程度上已被曲解，以致现代城市中常以构建大面积的景观绿地作为城市美化和更新的手段。

现代主义时期产生的城市环境，传统街区原本拥有的那种建筑与外部空间之间的集成连贯的关系已不复存在；反之，外部空间成了背景，而一幢幢的建筑俨然如浩大景观中的纪念物。景观在这种情况下也往往被设计成为图案感很强的形式，以此来加强建筑之间的关系。然而，这种关系却不是立体或多向度，因而显得单薄而疏离。与此同时，为进一步补偿人们在心理上对自然的渴求，现代主义城市中出现了各类集中绿地（如公园，休憩空间等）。当大面积公园绿地呈现无人使用状态，且无法满足人们社会活动的需求时，人们不得不作出反思，现代主义规划和现代主义建筑的某些理念是不是一种处理建筑与景观关系的理想模式。

In light of modernist planning principles, urban development was inclined to treat buildings as isolated objects situated within landscape rather than being an integral part of the spatial system comprising streets, squares and other viable open spaces. Consequently, there emerged a large number of anti-spaces, lacking in definition and shape in modern city centres. Within this kind of urban environment, buildings looked like monuments, with spaces between them being vast and formless and lacking a coherent connection. Indeed, the Master of modernist architecture, Le Corbusier once advocated that nature was to be spiritually enjoyed through the sun, sky, grass and trees. However, this argument somehow was misinterpreted, resulting in large areas of green landscape as a means of beautification and urban renewal.

In modernist urban developments, it was easy to find that the compact and coherent figure-ground relationship as shown in traditional towns did not exist any longer. Instead, the external environment had become the ground, with building blocks appearing as monuments. In this case, landscape was always designed with strong forms in order to establish a relationship between buildings. However, this kind of relationship was not multi-dimensional, but weak and alienated. Meanwhile, to further compensate the psychological needs of people for nature, the exceptional abundance of nature,such as city parks were provided inside the city centres. While a large number of open spaces were found to be ineffectively used, it was imperative for urban or architecture designers to rethink the relationship between architecture and landscape, and to endeavor their utmost to create a meaningful and harmonious environment for our cities.

a. 毯式建筑

到了 20 世纪中期，现代主义趋于教条的机械决定论以及孤立看待建筑与自然景观的组合模式也开始受到质疑，其中以 Team X[1] 为代表的年轻建筑师群体反应最为激进。他们认为，现代主义所提倡的功能分区和彼此孤立、缺乏有机联系的建筑环境恰恰是导致城市活力丧失的原因。在实践中，他们着力抛弃 CIAM[2] 的功能理想主义。通过对传统城市形态的剖析，并在对城市结构的复杂性及可生长性的模型进行探索的基础上，他们提出一系列强调结构的连续性、有机性及整体性的空间结构模式，并最终和柯布西耶一起完成了被称为“毯式建筑”的实践（图 1-9）。毯式建筑形态的出现，标志着现代主义建筑学走向一个新的方向。建筑不再是凌驾于自然环境或城市环境之上的一种操作手段，而是成为城市或自然环境中的一部分。

毯式建筑有自己独特的操作方式及空间特色，其中 Team X 设计的阿姆斯特丹医院和勒•柯布西耶设计的

a. Mat Architecture

In the mid-twentieth century, modernist ideology and mechanical determinism were gradually deemed controversial. One of the most influential opponents was Team X[1]. Being composed by a group of young architects, they contended that it was precisely the principle that advocated zonings and isolated high density built environment which led to the lack of coherent links and the loss of urban vitality. In practice, they abandoned the functional idealism promoted by CIAM[2]. Based on the analysis of traditional urban morphology and the exploration of complexity and growth pattern of urban structures, they proposed a series of spatial patterns to highlight the structural continuity and flexibility, and ultimately generated 'Mat architecture' together with Le Corbusier. Mat architecture, as a symbol of modernist architecture toward a new direction, no longer stood out against the natural or urban environment, but functioned as an integrated part of it.

威尼斯医院是毯式建筑的两个代表性作品。威尼斯医院是柯布西耶暮年的代表作之一，可以说是他设计生涯中较为特殊的一个作品，反映出随着Team X对现代主义的批判，柯布西耶也对现代主义的一些主张进行反思，并在实践中进行了尝试。威尼斯医院的设计出发点是思考如何在一个特殊的城市环境，把建筑融入城市结构中，使建筑具有城市肌理的延续性。基于此理念，威尼斯医院最终被设计成由四个符合人体尺度的护理单元组成，柯布西耶通过在建筑内部引入威尼斯的步行街道、广场、运河和现存建筑等典型的城市元素，重塑城市的氛围，并使每一个空间片段都能产生与威尼斯相关的联想与体验。

早期的毯式建筑可以说是建立在一种理想结构主义之上，具有本质主义的倾向，体现建筑师们试图抛开传统建筑学一贯采用的几何形式语言的束缚，探索建筑空间下更丰富的生活模式，从而让建筑能突破形式，与环境更契合、更一体。

注释:

① Team X：以彼得·史密森夫妇为首的一个青年建筑师组织。他们因在CIAM第十次大会上公开倡导自己的主张，反对以功能主义、机械美学为基础的现代主义理论而得名，对20世纪下半叶欧洲的建筑思想发展产生了深远的影响。Team X出现了两种不同的运动：英国成员（艾莉森和彼得·史密森）的新野蛮主义和荷兰成员（阿尔多·凡·埃克和雅各布·巴凯马）的结构主义。

②“国际现代建筑协会”（简称CIAM）是1928年在瑞士成立的第一个由国际建筑师组成的非政府组织。1933年CIAM第四次会议通过了《雅典宪章》，标志着现代主义建筑在国际建筑界占统领地位。进入50年代，对于现代主义建筑持反思批评态度的建筑师开始增多；在1959年荷兰鹿特丹举行的CIAM第十一次会议上，由于新老两派建筑师的严重分歧，导致CIAM宣告解散。

Mat architecture had its own unique operating strategy and spatial forms. The Amsterdam Hospital designed by Team X and Le Corbusier's Venice Hospital are two typical examples. The Venice Hospital was considered as a special design work during Le Corbusier's older age, revealing the fact that upon Team X's critique on modernism, Le Corbusier also reflected on some advocates of modernism and experimented with something different in his practice. The design of Venice Hospital was initiated from the idea of integrating building structure into the unique city environment, so that architecture could be a continuity of the urban fabric of Venice. Eventually, the hospital was designed to comprise four nursing units conforming to human scales. Through incorporating a certain number of typical urban elements, including pedestrian networks, squares, canals and existing buildings into the design, Corbusier reshaped the city's atmosphere inside the hospital, stimulating the imagination and experience of each space in relation to Venice.

Mat architecture at its early stage was mainly based on an ideal structuralism and had a tendency towards essentialism, implying that the architects tried to put aside the traditional architectural language of geometric forms, but explored the possibility of creating a richer lifestyle inside the buildings. As a result, the design could break the tradition of merely creating forms, being more integrated with the environment.

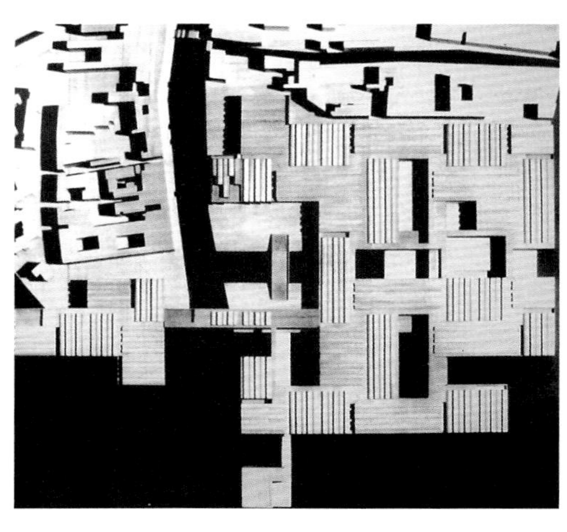

图1-9　柯布西耶设计的威尼斯医院是毯式建筑的代表作之一（1964年）

1.3.3 后现代主义背景下的自然化建筑理论

20 世纪 60 至 90 年代，建筑学发展进入后现代主义时期。后现代主义（Postmodernism）是一个从理论上难以精准下定论的概念，因为后现代主要理论家均反对以各种约定俗成的形式来界定其主义。若以单纯的历史发展角度来看，建筑学是最早出现后现代主义的领域之一，表现较为突出的是自 20 世纪 60 年代以来的建筑师，因反对现代主义盛行的国际风格（International Style）及其缺乏人文关注，而相继提出不同的理论与学说。

后现代主义是个提倡个性化的时代，面对趋于复杂性和多样性的现实世界，建筑学开始呈现不同风格并存、并不局限于单一规则的格局。而基于对现代主义进行反思的背景，自然与建筑环境关系问题再次成为公众探讨的焦点，各种理论与学说也相继涌现并明显地影响着设计实践。其中，亚历山大提出“城市不是一棵树”（图 1–10）的学说，反驳现代主义强调的理性和层次分明的观点，认为其并不能反映现实世界的复杂性和多样性；相反，城市和建筑环境应是类似自然景观的系统，呈现半网格结构及各部分之间的多向联系，从而展示自然世界丰富、多元及变化的特点。亚历山大理论的核心在于，一个好的设计并不是指一个机械系统中各个组成部分运作良好，而是指每一部分之间能相互激发、增长，从而形成一个相互关联的整体。因此，对于一个城市或建筑环境，我们不能局部看待其组成要素，而应把它们放置在一个相互依存的复杂系统中来考究。

1.3.3 Theoretical Foundation of N-architecture in Postmodernism

From the 1960s to 1990s, the development of architecture entered its postmodern era. Theoretically, postmodernism is a terminology difficult to define, because most postmodern theorists are opposed to any forms to define their doctrines. According to the historical development clues, architecture is one of the areas in which postmodernism first appeared. The most significant phenomenon was that the architects in the 1960s, being opposed to the prevailing international style and its lack of concerns with human needs, proposed various novel ideas and doctrines.

Postmodernism was an era to promote individualism. To facilitate the complexity and diversity of the real world, the architecture discipline also exhibited the coexistence of different styles. In light of the reflection and critique of modernism, the relationship between nature and built environment once again aroused intense public debates, resulting in the emergence of a number of theories and doctrines that significantly influenced the design practice.

Among them, 'A city is not a tree' advocated by Christopher Alexander criticizes the tree diagram that governed the planning of the time, a scheme where each part interacts with the whole through a hierarchical and pyramidal relationship, made up of subassemblies, collected in groups, that are connected in units of order that are gradually larger. In opposition to this deterministic diagram at the base of the artificial cities, Alexander proposes the model of the natural city. Such a city settles over time and is structured as a 'semi-lattice', an open structure where the parts are connected to each other by several orders of relationships, and the elements of a smaller scale may interact with others without being subjected to an inflexible hierarchy. This model enhances and facilitates multiple connections and informal levels of relationships between different orders of scale and significant interferences between the parts.

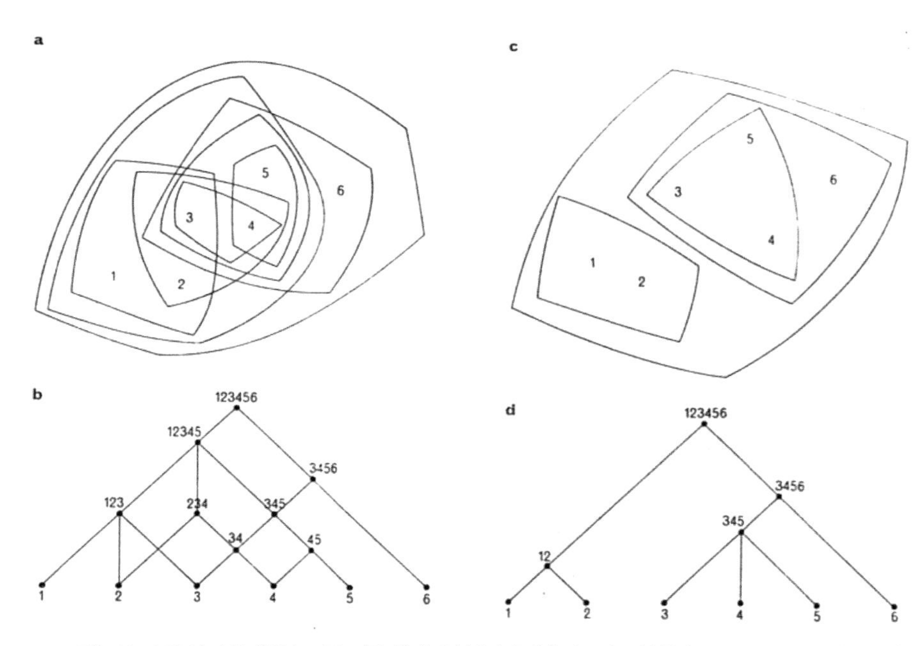

图1-10 亚历山大的“城市不是一棵树”指出城市空间结构应是半网格状（'semi-lattice'，如a、b）

a. 场域理论

而另一位理论家斯坦•艾伦(Stan Allen)基于现代主义后期出现的毯式建筑，于1985年提出“场域理论”。场域理论进一步模糊建筑与环境之间的界限，用以描述人员、物质、信息流相互影响下的一种空间现象或存在状态。这一理论的出现为建筑学开拓新的视角，使我们可以描述更为复杂自由的建筑存在状态。在一个特定的场域环境中，单元（包括建筑与自然景观的各种要素）是均等的，没有主次轻重之分，但相互之间存在内在的逻辑关系；而其外在的形态则处于一种不确定或适应的状态，一方面是内在逻辑的外在表现，而另一方面又承受外部环境的作用力（图1–11）。

场域理论可以说是西方建筑学理论发展的一个转折点，其拓展了建筑学的内涵，使建筑可以有另一种可能性，表达在自然世界中存在的一种状态。场域环境可以是任何形式或空间的母体，在遵循一定逻辑的基础上，将各种元素整合在一起。内在的多孔性和内部的连接关系决定了其边界模糊和松散的特点，并使其整体的形式和范围具有高度的流动性与适应性。场域理论自提出后，在设计实践中就产生非常丰富的表现形式；尤其当它与其他学科领域，比如生物群体行为领域、数字场域理论、非线性动力学等的研究相结合时，其研究与设计就从单一走向多元，从静态模拟走向动态。例如，20世纪80年代末，人工智能理论家克雷格•雷诺兹（Craig Reynolds）设计了一套计算机程式来模拟鸟类的群聚行为。结果发现，只要制定简单而准确的局部规则，鸟类每次都能自然而然成群。同

a. Field Theory

Another theorist, Stan Allen, with reference to the unique characteristics demonstrated by mat architecture in modernism, put forward field theory in 1985. Field theory further blurs the boundaries between architecture and nature, and describes a spatial phenomenon or state of existence resulting from the interaction of humans, materials and information flows. Field theory somehow broadens the boundary of architecture discipline by offering a new perspective, from which people can describe more complex and more flexible architectural conditions. In a particular field condition, various individual elements, including architecture and natural landscape, are equal, without stressing their primary and secondary levels; yet among them there exist logical connections. Incidentally, the form of the field is not determined but is in an adaptive state. The form is a manifestation of internal logic on the one hand, but on the other hand, it is subject to the external environmental forces.

Field theory is deemed a turning point for the development of Western architecture, owing to the fact that it extends the connotation of architecture, making architecture another possibility in the real world. Field conditions can be any forms of spaces following a certain logic and integrating various elements. The intrinsic connection and internal porosity determine the fuzzy boundaries and loose characters of fields, as well as enabling them to possess a liquid-

样，在模拟人的行为模式的研究当中，雷诺兹也发现类似的规则与现象，如麦加朝圣的人群（图1–12）。群聚显然是一种场域现象，而人或动物的群聚行为并不是简单的重复，也不是一种固定的模式，而是类似的结构或个体行为模式重复积累的结果。相比于鸟类，人群表现出更为复杂的动态和更少的固定模式之间的相互作用。雷诺兹两个针对生物群体行为的研究显示，场域其实是一种动态的表现形式，是局部性规则和逻辑起决定性作用。场域理论产生的意义在于，它使建筑学得以从传统的自上而下的控制转变为更为自由和适应性更强的自下而上的研究方法，使建筑与自然环境的关系更为整体和密切，并可以出现更复杂和多向度的形态。因此，场域理论是自然化建筑重要的理论基础之一。

like form and a high degree of adaptability. Based on field theory, architects across the world have created a variety of design works, especially when the theory is combined with other research disciplines, such as bio-field group behaviour, digital field theory, nonlinear dynamics, and its research has also increasingly evolved from single to multiple, and from static to dynamic simulation.

For example, at the end of the 1980s, artificial intelligence theorist Craig Reynolds designed a computer program to simulate the flocking behaviour of birds. It was found that, as long as simple and accurate local rules were made, the natural bird

图1–11 斯坦・艾伦的场域理论

flocks could be generated every time. Similarly, in the study of the simulation model of human behaviour, Reynolds found the same rules and phenomena, such as the Hajj crowds. Definitely, clustering was a field condition, and the flocking behavior of humans or animals was not a simple repetition nor a fixed type, but the result of the accumulation of a similar structure or individual behaviour patterns. In comparison with birds, humans exhibited a more complicated interplay of dynamic and less fixed patterns. Reynolds' two biological research projects on group behaviour have manifested that the manifestations of field conditions are actually dynamic, where localized rules and logics play a decisive role. Therefore, the significance of field theory lies in the fact that it allows architecture to transform from the traditional top-down control to a more flexible and more adaptive bottom-up approach, so that the relationship between architecture and nature appear more complex and multidimensional. In this sense, field theory is an important theoretical basis for N-architecture.

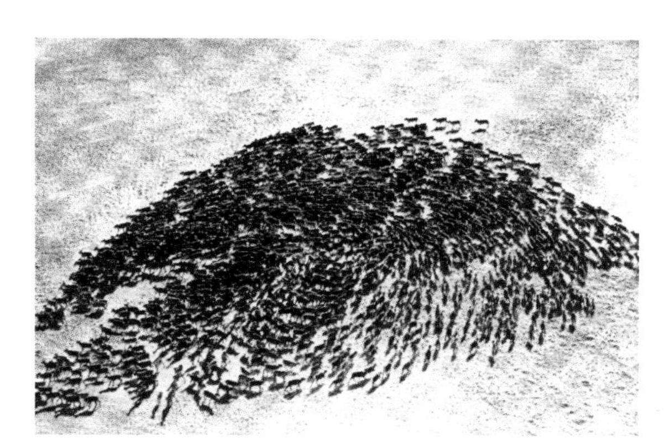

图1-12　羊群的群聚行为（上）和麦加朝圣的人群（下）

b. 混沌理论

当代另一位具有影响力的理论家是查尔斯·詹克斯。他在 *The Architecture of the Jumping Universe* 一书中，通过对复杂性科学的认识，总结了复杂性研究对建筑学发展的影响。他指出，有别于现代主义的简单与基于一个确定性机制的倾向，后现代主义建筑的核心是混沌和复杂性科学，并在此基础上，总结出后现代建筑的八个主要特征：

首先，建筑应尽量接近自然的语汇，这是把人类和自然缔结成一个整体的方式。通过扭转和折叠、波动和分形、自相似性等手段，设计不再是简单的重复。他指出，自然系统中一切事物，如水晶、骨头和云等，它们的物质形态不但遗传和延续了自然世界的复杂特性，同时每一种物质也拥有自己独特的自然性。因此，人类作为自然的一部分，是可以创造第二自然的（如自然化建筑）。其次，建筑应反映自然界的本质——自组织性、紧急及跳跃性。事物演进的基本方向就是走向复杂性，而这个过程可以是平缓过渡也可以是戏剧性突变，连续且跳跃，平滑且具“蝴蝶效应”。第三，建筑应反映系统组织的层次、多价值性、复杂性及边缘的模糊和混沌。另外，建筑应提倡多种性与多样性，建立自下而上的参与性系统，以保障事物之间最大程度的差异。而且，建筑的多样性可通过拼贴、激进的折衷和叠置来实现。另外，建筑体系应反映时间性和一些重要议题，其中包括生态紧迫和政治多元化的诉求等。再者，建筑应具有美学和理念上的双层价值体现，以实现多重关怀。最后，建筑应借助现代科学手段来发现自然界的规律，

b. Chaos Theory

Another influential contemporary theorist Charles Jencks, in his *The Architecture of the Jumping Universe*, summed up the impact of complexity science on the development of contemporary architecture. He stated that unlike modernist simplicity based on a deterministic mechanism, the core of postmodern architecture features chaos and complexity science. On this basis, he further summarized eight main features of postmodern architecture:

First, architecture should be close to nature and natural language, which is a holistic manner to tie people to nature. Design, therefore, with twists and folds, waves and fractals, is of self-similarity instead of exact repetition. When looking at the rhetoric of natural systems, such as crystals, bones and clouds, it could be found that these forms represent continuity with the natural world, and they might be supplemented with an additional source of creativity. We mankind thus can create a second nature (such as N-architecture).

Secondly, architecture is the representation of the basic cosmogonic truth-self-organization, emergence, and jumps to a higher or lower level. There is a basic direction of evolution towards increasing complexity, but it is attained through an oppositional process of gradual improvement and catastrophic change, continuity and jumps, smooth transitions, and the Butterfly Effect.

而非线性理论的发展为人类解决、表达和展现这种复杂性提供了更多可能。

詹克斯的混沌与复杂性观点突破传统建筑学的法则，指出建筑与自然系统本质上的一致性与共通性，从而为后现代建筑学发展提供了重要的理论依据（图1–13）。

Construction of buildings should also reflect the organizational depth, multivalence, complexity and the edge of chaos of the system. In addition, architecture should celebrate diversity, variety, and bottom-up participatory systems which maximize difference. And diversity can be supported by techniques such as collage, radical eclecticism and superposition.

Furthermore, architecture should acknowledge the time and its compelling agenda, which include the ecological imperative and political pluralism. It should have a double-coding of these concerns with aesthetic and conceptual codes. Finally, architecture must look to science, especially contemporary sciences, for disclosure of the Cosmic Code.

Through emphasizing the consistency and commonality between architecture and the natural system, Jencks' proposition has gone beyond the laws of traditional architecture, providing an important theoretical basis for the development of postmodern architecture.

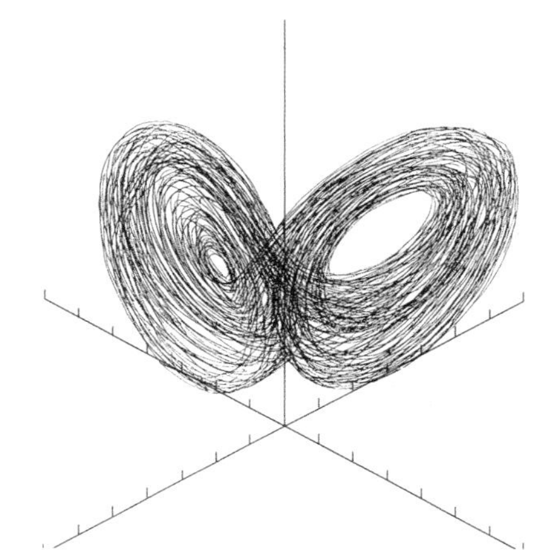

图1–13 查尔斯·詹克斯展示18世纪的Orrery（上）与20世纪的奇怪吸引子（下）并指出每个世界观的差异都是有启发性的。牛顿和现代主义者认为，自然如教学模式是可预测的，就像行星及卫星一样沿着他们预先设定的轨道运行。而对于后现代主义者，奇怪吸引子（由爱德华·洛伦茨发现）是一个更好的自然模型：天气总是自相似的，但却从来没有完全重复。它是由整体拉压力、温度和其他因素的相互作用而形成两个向心趋势，像两只“眼睛”，同时受到两个混沌吸引子的限制。而这种持续的综合作用折叠成朝向彼此的复发形状，总处于变化却又总是相似，可称之为“蝴蝶效应”。我们现在知道，宇宙是一个混沌的吸引场，而不是一个具有确定性的机制。

c. 分形理论

“分形理论”从另一角度为自然化建筑的实现奠定了理论基础。分形理论诞生于20世纪70年代中；顾名思义，分形理论是通过规律性法则来描述自然界中不规则事物的科学（图1-14）。自然界其实并不是如欧几米德般充满逻辑、秩序和理性，与之相对立的是，在自然界中其实找不到一个完全可用欧式几何描述的形式；例如，山不是圆锥形，植物也形态各异，这些都需要一种新的观点和视角来描述。在此情形下，芒德勃罗创立的分形几何学应运而生。分形是一种新的方法论与自然观，从分形的观点来看世界，可以发现，分形状态其实是自然界中普遍存在的状态之一，整个世界正是以分形的方式在不断演变与进化的。

基于分形的观点，不难发现，现代主义建筑的缺失就在于其过分强调建筑的自我导向功能及整体感，忽略尺度层次和细节表现，导致建筑冷漠及与环境格格不入的特性。虽然人们对建筑的感知既来自整体也来自细节，然而这两者并不矛盾。基于分形理论，细节通过在整体下的分形来表现，而分形下的部分具有丰富整体层次及体现细节的作用。因此，基于分形理论指导下的设计可以更贴近自然形态，尺度上更趋向于自然环境。在传统的建筑和手工艺品中，分形几何形是常被采用的设计手法之一。

c. Fractal Theory

Fractal theory, from another angle, laid the theoretical foundation for the realization of N-architecture. Fractal theory was proposed by Benoit B. Mandelbrot in the 1970s. As its name suggests, fractal theory is a science to describe the regularity rules underlying irregular things of nature. In truth, nature is not full of logic, order, and reason as Euclidean described. By contrast, we could not find a form which can be precisely described by Euclidean geometry. For example, mountains are not conical, and plants come in various shapes, which requires a new theory to describe and represent them. In this circumstance, fractal theory emerged. According to Mandelbrot, a fractal is a rough or fragmented geometric shape that can be subdivided into parts, each of which (at least approximately) is a reduced-size copy of the whole. Therefore, fractal theory is a new methodology and perspective, from which the most common state of the real world and the way the world evolves are able to be described and understood.

In light of fractal theory, it can be inferred that modernist architecture put too much emphasis on self-directed functions and the entire form of buildings, ignoring the various orders of scale and levels of detail, subsequently leading to an apathy and incompatible built environment. In principle, people's perception of buildings derives from both overall form and details, but they are not contradictory. Being controlled by the entirety of buildings, a fractal plays an important role in expressing architectural richness and refinement. Therefore, design works with the inspiration of fractals would be closer to the natural language and more compatible with the natural environment. In traditional architecture and handicrafts, fractal geometry is one of the commonly used design techniques.

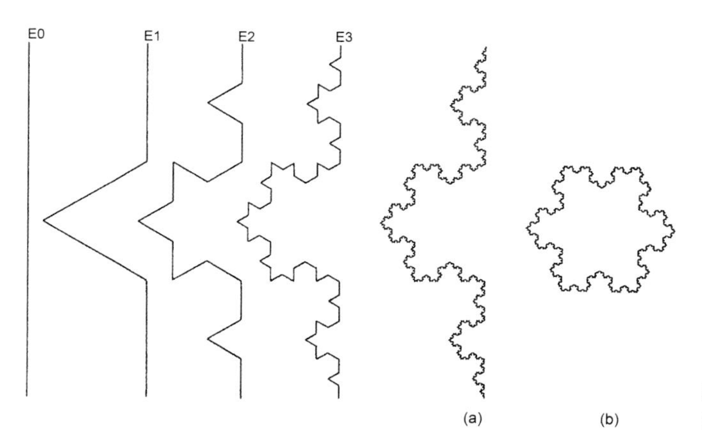

(a) 冯・科奇曲线的建立
(b) 三个冯・科奇曲线组成雪花曲线

图1-14　分形几何形

1.4 小结

综上所述，建筑与自然景观的关系一直是伴随建筑学发展的一个重要话题。从远古的原生态居住原型，到现代主义的孤立与机械化，再到 20 世纪后期的多元、复杂、混沌建筑作品的出现，都具有一定理论前提条件。而本书所提出的自然化建筑作为其中一种建构形式，一种书写场地环境关系的新型建筑策略，其一方面得益于建筑学领域的理论发展，如对当代建筑学发展具有相当影响力的场域理论及詹克斯的混沌和复杂性科学等；而另一方面，科学技术的发展以及当代景观学、建筑学、城市形态学科之间的交叉融合，也使自然化建筑的逐步成形与独树一帜成为可能。

1.4 Epilogue

The relationship between architecture and natural landscape is an important topic accompanied with the development of architecture. From the original living prototype in ancient times to modernist doctrine of isolation and mechanization, and to pluralism, complexity science and chaos occurred in the late twentieth century, they all have certain theoretical preconditions. Inevitably, N-architecture discussed in this book considered as a form of creation, a strategy to establish the relationship between context and architecture, benefits from the theoretical development in the field of architecture, including Stan Allan's Field Theory and Jencks' Chaos and Complexity Science on the one hand; on the other hand, the magnificence of science and technology, as well as the interdisciplinary approach among landscape, architecture, and urban planning, have also stimulated the evolution and realization of N-architecture.

02 形态学分析：自然化建筑的概念模型

Morphological Analysis: Conceptual Model of N-architecture

伴随着建筑学的发展历程，自然化建筑已有不同的形态表现。从远古的居住原型到当代的参数化设计，自然景观都成为其中一种要素或源动力，以不同形式参与建筑空间的建构。

During the development of architecture, there have been different forms of naturalized architecture. From the original living prototypes in ancient times to the contemporary parametric design, natural landscape was always deemed a driving force or an important element, participating in the spatial articulation in a variety of manners.

2.1 隐喻和象征

2.1 Metaphor and Symbolism of Nature

自然化建筑最直接的表现形式是模拟自然界的地形地貌特征。例如，建筑取山、石、洞、甚至是水的形态，隐喻自然。在这类建筑的表现当中，自然界的物质形态经过抽象与演化，形成特定的建筑语言。Reiser+Umemoto 于 1992 年设计的“水花园”（Water Garden）就是其中一个比较典型的设计。水花园虽是景观建筑，但却是很好地展示模拟自然的设计手段运用的例子。水花园的业主是俄亥俄州立大学建筑学院教授杰弗瑞 • 肯尼斯（Jeffery Kipnis）及其妻子贝弗利 • 斯蒂芬（Beverly Stephen）；他们希望打破基于稳定几何形态的设计传统，藉此创造一个动态的建筑景观，以实现与自然比较彻底的互动。

因此，水花园设计概念的产生源于对传统设计观念的一种批判性思考。设计师（R+U）认为，西方建筑（从古典主义到现代主义）都存在一种根深蒂固的观念，也就是在固定功能和传统的表象背后存在某些一成不变的内容，而这些内容最终被演化成简单化的几何模式语言或约定俗成的建筑学类型；倘若我们能突破这些固有观念，把建筑

The most obvious manifestation of N-architecture is the building that takes its imagery from the natural geological features. For example, some buildings are actually idealized mountains, rocks, caves, and even water. In this case, architecture is considered as a metaphor or symbolism of nature, with the natural phenomena being transformed into particular design language. By means of abstraction and optimization, architecture establishes its inherent connection with nature.

The Water Garden designed by Reiser + Umemoto in 1992 is typically a metaphorical example. The Water Garden belongs to landscape design, but reveals a successful implementation of the concept of metaphor and symbolism. The clients, the architecture professor of the University of Ohio, Jeffery Kipnis and his wife Beverly Stephen, attempted to create a type of landscape architecture, whose dynamic form could interact with nature, contrasting with what they considered to be the staid geometry and passivity of historic gardens.

从静态及被动的思维惯性中解放出来，建筑设计可以进入一个全新的领域。基于此信念，设计师在基地内引入一组堤坝系统，形如相互交错的一双手，有意识地形成一种形态上的不稳定感。原有硬质庭院被延伸至土丘上，并通过翘曲、凹陷、皱褶等变异手段，形成类波浪纹的景观体系。然后，在波浪纹理中不规则地种植草和水生植物，而这些植物可随着季节变化而呈现景观上的差异。四道径流渠道从护堤流入庭院，并通过一套水泵系统和底层横向混凝土板体系来控制水流，从而使庭院内的水位时高时低，景致也随之产生变化。这个设计有趣的地方在于，整个花园创建了与自然有趣的关系，自然形态（如水）不仅被抽象化，同时时间维度也被融入到设计当中，从而使花园具有类似自然的生命力。水上花园很好地模拟水流冲刷河岸并形成旋涡的场景，并以隐喻自然的方式来构建复杂的景观形态（图2-1）。

As a result, the design concept was initiated from a critical thinking of design traditions. The architect believed that there was a deep-rooted tradition in Western architecture (from classicism to modernism). That is, there exists a permanent and unchanging essence behind the world of appearances, and such essences finally universalize themselves in fixed simple geometries and timeless typologies. If architectural design can be liberated from the static and passive inertia of thinking, the practice can enter into a new era. Based on this belief, the architect then introduced a berm system resembling interlocked fingers across the backyard. This was the first act of a deliberate production of instability. Following this, the existing concrete patio was extended over the mounds. After warping,

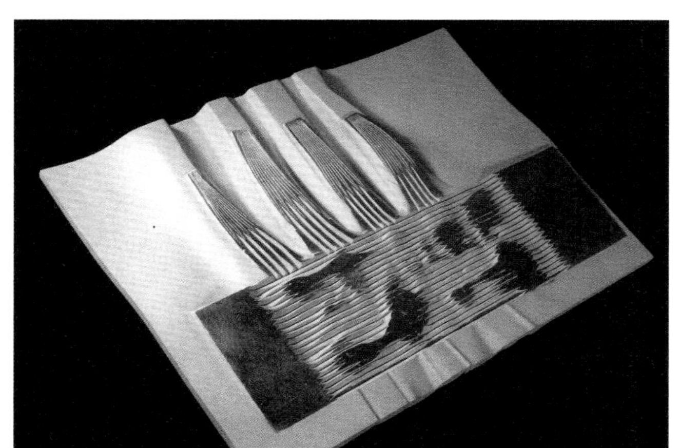

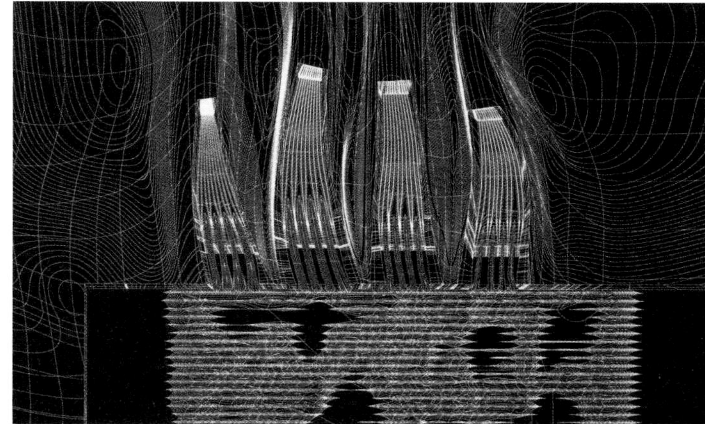

图2-1　水花园景观设计（R+U）

假如说水花园的设计是对自然比较直接的模仿，而在隐喻上倾向于抽象化表达的建筑师则有高迪（Gaudi）。高迪通常被认为是加泰罗尼亚现代主义大师（Catalan Modernism），然而他的作品却超越任何一种风格或分类，独树一帜。其作品具有丰富的想象力，充分体现自然启示的特点。高迪的设计灵感主要来自他对自然表现形式的深入研究，并把这些自由生动的形式用建筑手段展现出来；例如，他成功地把自然界丰富的元素诠释成建筑形态语言，形成双曲抛物面、双曲面、螺旋等。高迪发现这些复杂几何形态其实广泛地存在自然界当中，如蒲草、芦苇、骨骼；这些几何形态同时兼备功能及美学的双重价值，而这恰恰是建筑与自然共通的属性，而这种一致性也是致使他醉心于研究自然的原因之一。

高迪把他对自然的研究及得到的启示广泛地运用于个人的建筑创作当中，并在他的代表作之一的圣家堂（Sagrada Familia）得到充分体现。圣家堂是西班牙巴塞罗那一座天主教堂，其高耸与独特的建筑设计，使得它成为巴塞罗那最为人知的观光景点之一（图 2-2）。圣家堂自 1882 年开始修建，由于资金的来源主要靠个人捐款，因此捐款的多少直接影响到工程进度的快慢。圣家堂至今还未完工，但却是世界上唯一一座尚未完工就被列为世界遗产的建筑物。高迪设想把整个教堂设计成一个类似森林的结构，由一系列树状柱支撑着相互交织的双曲面。同时，他还创建了不需要扶壁就可以完美支持牵引力的结构体系，先锋性地实现建筑的理性化、结构化与逻辑化，并与此同时形成自身独特、简朴、优美和令人赞叹的建筑风格。

denting, wrinkling and other means of variation, a wave-like pattern was created. Being irregularly planted with grasses and some aquatic plants, the garden finally acquired natural variation and diversity, with its scenery changing with seasons. In addition, four runoff channels were projected into the courtyard from the berms, with the level of water falling and rising periodically, which was mainly controlled by either an optional pump system or the striated concrete slabs underneath.

The significance of the garden design is that it has created an interesting relationship with the natural environment. The geological features (such as water) was not only abstracted into a design language, but the time dimension was also engaged in the process of the formal generation. As a result, the garden has gained its natural vitality. The Water Garden is a good simulation of waves sweeping onto a beach and forming vortexes, demonstrating that metaphor and symbolism are effective measures to generate complex architectural forms.

While the Water Garden tends to imitate nature in a straightforward way, the world's famous architect Gaudi was inclined to adopt abstract expressions of nature in his design works. Gaudi is usually considered as the greatest master of Catalan modernist architecture; however, his works have gone beyond any style or classification, because they are so imaginative and dramatically full of inspirations from nature. Through studying the organic and anarchic

采用类似的手法进行自然化建筑创作的还有西班牙另一建筑师安瑞科•米拉莱斯（Enric Miralles）。在其设计的伊瓜拉达公墓 (Igualada Cemetery) 当中，有别于在场地内建立一种新秩序的做法，他反其道而行之，通过建筑手段来强化现有的自然景观序列，从而使建筑环境完全融入自然环境中（图 2-3）。胡安•何塞（Juan Jose Lahuerta）在描述伊瓜拉达公墓时曾道：“在这个疑似破碎的环境中，或者说在一些片刻，土地以某种方式裂变，形成让人窒息、不忍目睹的情景；凹凸不平的石块和锈迹斑斑的物件咬合着……”在这种场景中，我们根本无法区分建筑和自然景观，这是一个有悖于我们惯常理解的建筑环境，是建筑完全被自然化的一种表现。

geometric forms of nature thoroughly, Gaudi accumulated countless ideas and expressed the understanding of these forms in his design works subsequently. For example, his research on nature was always translated into ruled geometrical forms, such as the hyperbolic paraboloid, the hyperboloid, the helicoid and the cone, which are commonly found in the natural world, for instance, in rushes, reeds and bones. Gaudi used to say that there was no better structure than the trunk of a tree or a human skeleton. These forms are at the same time functional and aesthetic, and thus are good to be adopted as structural forms in architecture.

The inspiration of nature is widely revealed in Gaudi's architecture, notably in one of his masterpieces, Sagrada Familia. Sagrada Familia is a Catholic church in Barcelona, famous for its unique architectural style. Although Sagrada Familia began to be built in 1882, due to its funding mainly relying on private donations, which might have affected the progress of the project directly, until today it has not been completed. Notwithstanding its incompleteness in construction, it was listed as a World Heritage Architecture by UNESCO in 1984.

Gaudi conceived the design of the church as a forest structure, comprising a series of tree-like columns, supporting the intertwined hyperboloids. In this project, Gaudi created a pioneering structure to support the transaction without buttresses, perfectly realizing functional, structural and logical integration within the architecture, and further formulating his own unique, simple and elegant architectural style.

Another Spanish architect Enric Miralles employed a similar design approach as Gaudi to create N-architecture. When designing the Igualada Cemetery in Spain, rather than establish a new architectural order on the site, Enric strengthened the existing natural landscape pattern with architectural forms, so that the artificial environment could be fully infilled into the landscape. In describing Miralles' Igualada Cemetery, the famous architectural critic,

图2-2 巴塞罗那的圣家堂是高迪的代表作之一

Juan Jose Lahuerta said, 'In this broken landscape or in these times, the earth, being dragged this way, held up this way, and split in this fashion, becomes unbearable. This is the most terrible erotic landscape, Iris death. Pointy teeth sprout from this cut, jagged stones teeth with already rusty devices...' In this scenario, one could never distinguish the architecture from its natural environment. Irrespective of being contrary to our conventional understanding, this is a successful case to manifest how architecture can be completely naturalized.

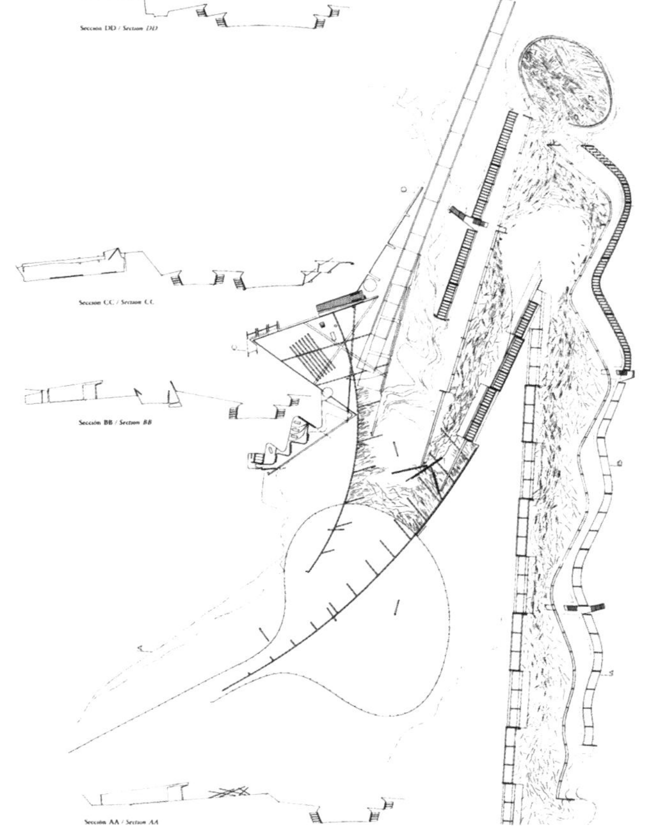

图2-3　伊瓜拉达公墓

2.2 融合自然

2.2.1 山地建筑

山地建筑是融合自然比较突出的表现形式。山地建筑一般指的是在陡坡上结合地貌特征，依照坡地差异进行建筑布局，追求错落有致的效果，从而使建筑与山地自然景观融合协调，避免高切坡现象的建筑形态。

日本著名建筑师安藤忠雄认为，建筑设计其实最终都可归结为如何回应场地需求的问题。设计的最终成果是自然界的逻辑与建筑逻辑矛盾冲突而又相互并存的状态，是一种双方妥协与融入的关系。因此，在山地建筑设计当中，设计师们常把建筑嵌入地形中，并以退台的方式来处理建筑体量以达到与环境融合的目的；同时以此为手段，为使用者提供一种在身体及心理上都具独特性的场景体验。

山地作为一种自然形态，广泛存在于人类的生存环境当中。基于开拓生存空间以及回归自然的需求，山地坡地成了人类社会空间发展的重要资源。因此，山地建筑在世界各地并不鲜见，其中一个比较著名的案例是位于日本神户的六甲山集合住宅。六甲山集合住宅是安藤忠雄于 1983 年设计的代表作品之一，背依六甲山，前方俯瞰神户市区和大阪湾；建筑群体沿山势呈台阶状排列，嵌入山坡谷地中。场地的坡度高达 60 度，一般认为在如此陡峭的山坡上建筑是无法生根的。然而，安藤因地制宜，采取把斜面削平之后将建筑嵌入的方法，顺应山势，将建筑的垂直立面改造成呈阶梯状排布的倾斜立面；如此一来，限

2.2 Merging with Nature

2.2.1 Fitting with Slopes

Another prominent manifestation of N-architecture is fitting buildings into the slopes as mountain architecture does. So, what is mountain architecture? Normally, it refers to the buildings located along a steep slope and whose layout is carefully configured by taking into account the specific geomorphologic features. To avoid the high-cut slope phenomenon, buildings are designed to conform to the terrain, with their volumes scattered in a multi-level pattern.

Tadao Ando, the famous Japanese architect believed that architecture is ultimately a question of how one responds to the demands made by the land. The aim of architecture is therefore the creation of a new environment where the logic of nature and the logic of architecture are in fierce conflict yet co-exist. In this sense, when designing mountain architecture, most architects are inclined to arrange building volume into steps so as to be embedded within the environment in addition to providing inhabitants with unique experiences both physically and psychologically. As a typical geographical form, mountain slopes are widely present in our living environment. To explore more living spaces and to satisfy the demand for easy access to nature, mountain slopes have become an important spatial resource for social development. This may help ex-

高（按日本法规山地建筑不得高于两层）和视线遮蔽的问题迎刃而解。

六甲山集合住宅的建筑整体是基于一个 5.2 平方米的网格系统来生成，由三个建筑组群构成。每个组群大约为五个网格的面积，均质网格因应地形变化，布局上产生角度扭转和错位，演变成相对复杂的体系，从而在场地内建立新的建筑秩序。而这种变异同时也允许东向阳光进入建筑组群内部，形成空间上的光影特色。建筑群体中央配置一部楼梯，楼梯穿越三个单元组群；而每个单元各设有电梯，中央楼梯悬浮于中间标高的平面之上，并与三个组群的入口相结合。每个组群单元被切分成东西两翼，两翼之间的间隙不仅为住户提供交往空间，同时也是解决通风与采光等技术问题的重要措施。

六甲山集合住宅很好地展示了建筑如何与山形地貌相结合的设计思路。剖面上，它遵循了陡坡的斜率来布置空间，而平面上却采用相对规整对称的逻辑来组织控制；因地形原因而产生的变异也带来户型设计的变化。立面设计上，混凝土构件形成的立体网格形态与周边绿色植物交织在一起，既强调人工技艺，也凸显自然和谐，因而产生强烈的建筑美感（图 2-4）。

plain why mountain architecture is widely found across the world.

The Rokko Collective Housing located in Kobe of Japan is well-known for its successful dealing with slopes. Rokko Housing was designed by Tadao Ando in 1983. Being situated along the Rokko Mountain and overlooking downtown Kobe and Osaka Bay, the entire group of buildings were arranged in stepped forms, embedded into the valley. In fact, the slope of the mountain was up to 60 degrees, which normally was considered unsuitable for any construction. Nevertheless, Ando transformed the slope into a ladder-like terrain, fitting the buildings into the new terrain and further conforming to the topography by designing an inclined building façade. As a result, both the height limit (according to Japanese Building Regulations, mountain buildings should not be higher than two floors) and the obscured visibility problems were successfully solved.

The structure of Rokko Housing as a whole was based on a 5.2m^2 module to generate a grid system. The whole structure is composed of three conjunctive units, each unit being assembled on a base plane five times the area of the grid. To respond to the fierce slope conditions, the uniform grid was transformed into a relatively complex system so as to establish a new architectural logic. As a result, this variation allows sunlight into the buildings on the one hand, and endows the interior spaces with a unique light characteristic on the other. A central stairway is raised directly through the three units, and each unit is equipped with elevators from its basement. This stairway is suspended at the intermediate level, in combination with the entrances of the three units. Each unit is cut into east and west wings with a gap running from the north to the south. The gap not only provides communal spaces for the occupants, but also solves ventilation, lighting and other technical problems.

图2-4a　安藤忠雄的六甲山集合住宅

Rokko Housing has set a good example of fitting buildings into mountain slopes. In section, it follows the steep slope, with its plan structure being symmetrical. Resulting from the impact of irregular topography, variation is created in dwelling unit types due to the asymmetry occurring in its structural forms. As to its elevation design, the three-dimensional concrete grid structure is intertwined with the surrounding greenery, emphasizing a harmonious relationship between artifact and nature, consequently leading to a strong sense of beauty.

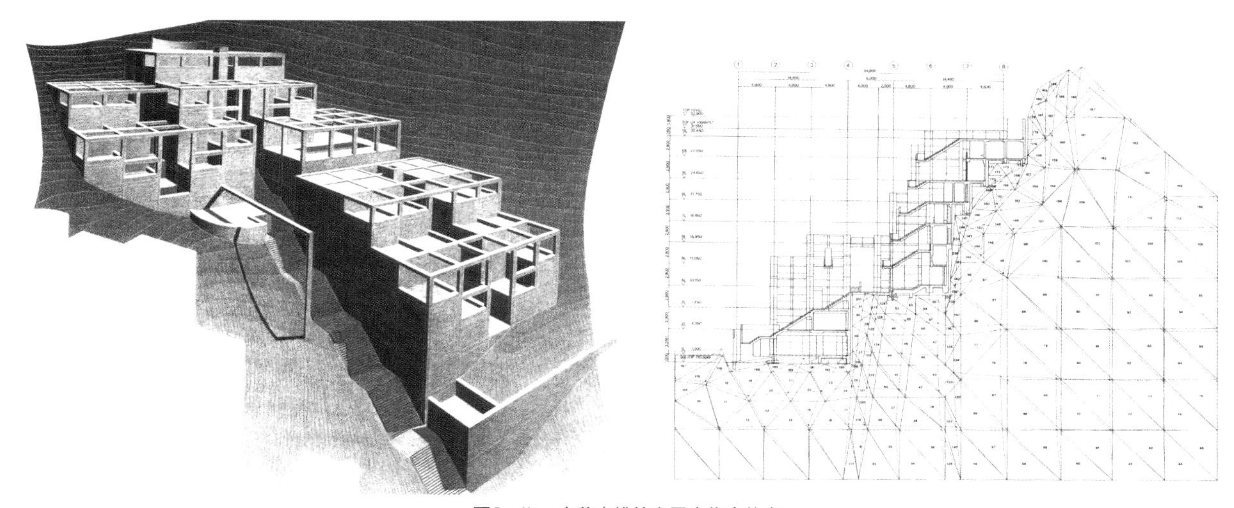

图2-4b　安藤忠雄的六甲山集合住宅

2.2.2 覆土建筑

覆土建筑是另一种最能表现建筑自然化的建筑形态。有别于山地建筑，覆土建筑是通过彻底消隐来达到与自然融合的目的。覆土建筑是一门既古老又新兴的建筑类型；在远古时期，人类就开始以穴居或石窟等生活模式来适应自然，以达到生存的目的。可以说，覆土建筑是人类最原始的居住形态之一。然而，在科学技术高度发达的今天，这种建筑形态也给当代建筑学发展带来相当多的启示，在巧妙利用与顺应自然环境方面做出了榜样。

覆土建筑指的是完全或局部隐藏在地里的建筑物；然而，它并不是指建筑物简单地被掩埋在地下，而是在其中隐含着许多建筑智慧。在世界各地留存的覆土建筑案例中可以发现，人类在与自然互动妥协的进程中，无论在技术上还是在功能上都已经发展出一系列适应性设计策略。尤其是在世界各地的乡土建筑当中，覆土建筑是其中一种适应自然的基本居住形态；例如，我国黄土高原地区的窑洞，就是这种适应性生活模式的典范。而到了当代，尤其是自20 世纪 60 年代开始，人们开始意识到将建筑物融入地形地貌之中的若干好处，除了保护地面的自然景观资源，减少对环境的干预和破坏，也是实现建筑与自然彻底融合的一种手段。因此，在设计领域已有不少建筑师采用覆土的策略来进行建筑创作。

当代覆土建筑的代表建筑师有埃米利奥（Emilio Ambasz），他早期的设计作品就广泛采用消隐的设计手法。堤坝居所是他其中一个具有里程碑意义的作品。该建

2.2.2 Earth-sheltered Structure

Earth-sheltered structures are another morphological manifestation of N-architecture. In comparison with mountain architecture fitting into the terrain, earth-sheltered structures merge with nature in a more straightforward way. The history of earth-sheltered architecture can be traced back to ancient times when human beings began to live in caves or grottoes to protect themselves against natural disasters. Although it was one of the primary living prototypes in human society, earth-sheltered structures have increasingly played a pivotal role in inspiring contemporary architectural designs in terms of their high level of adaptability and appropriate utilization of natural resources.

In the main, an earth-sheltered structure refers to buildings placed completely or partially underground, but not just simply buried. Rather, they comprise a great deal of wisdom, embodying a series of adaptive design strategies generated from the long-term interaction with nature by human beings. It is noticeable that earth-sheltered architecture can be found worldwide, notably in vernacular architecture, including the cave dwellings in the north of China.

Since the early 1960s, architects across the world have recognized that there are a considerable number of advantages to placing living spaces into the earth. In addition to preserving natural landscape and reducing interference with the environment, it

筑如此安静祥和地隐藏在自然当中，住宅的屋顶就是地表面，墙面犹如堤坝，建筑完全成为自然的一部分。除此以外，住宅也借助消隐于地表之中来实现一系列的生态化设计，从而减少70%的能源消耗；例如，自然通风代替空调系统、被动式太阳能利用等。南向大面积的玻璃窗设计，也使室内空间获得足够的光线。

堤坝居所的起居室采用的是开放式空间格局，可以为各类家庭活动提供灵活多变的场所。卧室空间主要围绕一内庭院采光，内庭院同时也是地下层标高的入口空间。两个大尺度的护堤结构把住宅与周边街道分隔开来，并形成地面标高的入口。护堤结构上覆盖太阳能板，以达到利用太阳能的目的。从这个设计作品我们可以窥见，堤坝居所不仅消隐在自然景观当中，空间设计上也实现与周围环境的一体化。覆土并不是简单的掩埋，而是利用剖面上的切角关系，巧妙地解决建筑的采光、通风等问题，并实现建筑环境的高度自然化（图 2-5）。

图2-5a　埃米利奥的堤坝居所

enables architecture to establish an intimate relationship with the landscape. In this sense, numerous architects around the world are inclined to employ earth-sheltered strategies to solve particular design problems.

One of the outstanding representative architects is Emilio Ambasz, whose early works mainly featured underground forms. Among them, the Berm House was a landmark project. The house is quietly hidden from the environment, at peace with nature. Its roof and walls facing the north are insulated by means of gently terraced earth berms, turning the house into an integral, unobtrusive part of the landscape. Except for this, by achieving cross-ventilation the house eliminates the need for air-conditioning. More than that, passive solar energy, complemented by solar collectors, has helped reduce the energy consumption of the house by 70% in comparison with conventional designs. The large windows facing southward allow for sufficient sunlight to project into the interior spaces, creating a cozy living atmosphere for the household.

The house mainly encompasses a spacious living room, a large dining area, a studio or library, master bedroom and guest bedrooms. The living area, placed between the parents' and the children's quarters, provides privacy to both, its open plan layout offering a flexible space for holding different family activities. The main living spaces, set around an inner courtyard, also function as a graceful entr-

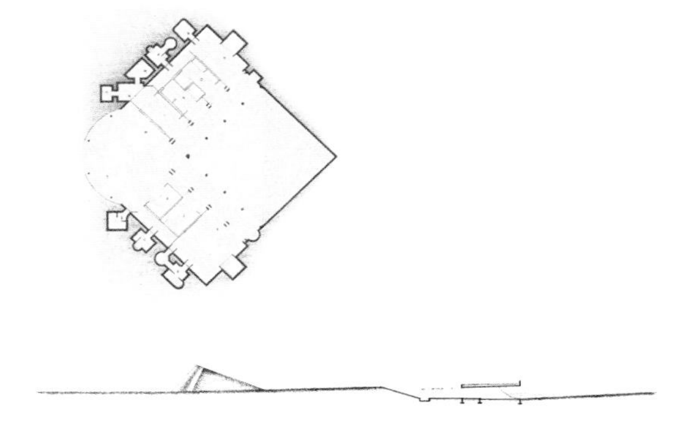

图2-5b　埃米利奥的堤坝居所平面、剖面图

2.2.3　有机建筑

著名建筑大师弗兰克 • 劳埃德 • 赖特认为：“大自然不仅为建筑提供了场地、材料等无尽的资源，也使建筑形式的持续发展成为可能。虽然近一个世纪以来，我们的设计实践已背离建筑这种自然本质，转而从书本经典中寻找灵感，或者从一些死板的公式原理中寻找答案。然而，自然的启示却是永恒的，取之不尽、用之不竭，它比我们所能期望与索求的要长阔高深。”这段文字道出他对自然与建筑关系的理解，并成为他创立“有机建筑论”的出发点及思想基础。

有机建筑准确来说是一种建筑哲学思想，由赖特于20世纪初提出。该思想主张建筑无论是外在还是内部结构都要与环境相得益彰，而这种建筑与自然的结合关系就是有机建筑论的核心。赖特指出，有机建筑同时也应包含整

ance at the underground level. Being covered with solar panels, two large earth berms shield the house from the surrounding streets, framing another entrance on the ground level.

From this project, people can get a glimpse of the implementation of the earth-sheltered design concept. Architecture is not simply placed underground, but achieves natural lighting and ventilation through an intelligent cross-sectional design. By means of these measures, the naturalization of living spaces is perfectly achieved.

2.2.3 Organic Architecture

Frank Lloyd Wright stated, 'Primarily, nature furnishes materials for architectural motifs, out of which the architectural forms as we know them today have been developed, and although our practice for centuries has been for the most part to turn from her, seeking inspiration in books and adhering slavishly to dead formulae, her wealth of suggestion is inexhaustible, and her richness is greater than any man's desire.'

The above paragraph not only reflects Wright's profound understanding of the relationship between nature and architecture, but also laid the ideological foundation for his organic architecture. Precisely, organic architecture is a theory proposed in the early twentieth century, asserting that both external and internal structures should harmonize with the

体与局部的和谐关系，并与自然界生命历程，如初生、成长、死亡，存在内在的一致性。有机建筑讲求结构体系每一部分要有自己的特点，并能在建筑整体中凸显自己的个性。结构体系之间是容纳各种活动的功能空间，彻底改变了传统的组织建筑空间的方式。在赖特看来，每一个建筑都是神圣的，要与它所处的环境息息相关，并成为其中的一部分。

有机建筑的代表杰作是赖特于 1934 年设计的“流水别墅”。在这座别墅的设计当中，赖特把人完全置于自然环境当中，峡谷、树木、花草都成为与人互动的元素，只要置身于别墅当中，就能马上体验到自然的荣美。通过这种方式，自然成了别墅主人生活的一部分。别墅在主要楼层提供了三个方向的景观，延伸出两个露台：一个在瀑布上游，另一个在岩石和瀑布之上。垂直方向的结构元素采用当地石材砌筑，强调自然凹凸的肌理，并产生强烈的雕塑感；而水平向构件以混凝土浇筑，地面则铺设与墙面一致的石材。混凝土楼板悬浮在瀑布之上，通过起居室的一部楼梯可直接到达瀑布。在主要楼层的上方，也就是第三层，卧室区延伸出来形成宽阔的露台，强调建筑整体形态的水平构成特色，并形成自然、舒展、随意的建筑风格。在流水别墅设计中，我们发现有机建筑通过材料、结构、空间与形式，使人与自然悠然共存，呈现天人合一的最高境界，这也是它成为世界无与伦比的著名建筑的主要原因。流水别墅向我们展示，自然化建筑可以如覆土建筑般无形与消隐，也可以是人工技艺的高度体现（图 2-6）。

environment through organic forms. In other words, it is the connection between nature and architecture that forms the core belief of the theory. Organic architecture subsumes the harmonic relationship between the whole and the parts, and is asserted to contain the process of natural life, including birth, growth, and death. Each part of the structure should have its own identity, but at the same time being an integral part of the whole. Normally, programmes are accommodated among different structures. This somehow has changed the traditional way of organizing spaces. Wright further believed that each building can be viewed as something sacred, closely linked with its environment and part of it.

The Kaufman House (also known as Falling Water Villa), one of Wright's masterpieces designed in 1934, is a leading project of organic architecture. In this villa, Wright attempted to put the occupants into an environment that could allow them to interact with the glen, the trees, the foliage and wild flowers entirely. As a result, wherever they were within the building, the glory of the natural landscape was accentuated, brought in, becoming a component of their daily life.

The main floor of the villa provided views in three directions, with two terraces extending over the landscape; one terrace opened upstream, and the other projected over the rocks and cascades. While the vertical structural elements of the house used local stone masonry to highlight the uneven texture and strong sense of sculpture, the horizontal elements were equipped with poured concrete, with its overall floors being paved with stones similar to the walls.

Moreover, the reinforced-concrete slabs were cantilevered from the rock band in order to enable the house to float over the stream. From the living room, one could step directly down from a suspended stairway to the stream. Immediately above, on the third level, terraces were extended from the sleeping quarters, emphasizing the horizontal characteristics of the house. As a result, a natural, flexible, graceful architectural style was formulated. From this design, one can be aware that through the interplay of materials, structure, space and forms, human and nature can coexist in a leisurely manner, and a high quality of real-time experience can also be generated. In this sense, Falling Water Villa has revealed us a fact that N-architecture can be as invisible as earth-sheltered structures, but also can be a celebration of artificial techniques and skills.

图2-6　赖特的流水别墅

2.3 图底关系的模糊

上章讨论的图底原理在城市学科领域常用于描述城市实体与虚体空间之间的形态与比例关系，是自然化建筑其中一个思考方向或设计逻辑。图底原理在建筑环境中的运用，主要用以描述室内与室外空间的关系，而图底关系模糊也主要指两者在界面上的模糊。也就是说，建筑内部与外部环境的关系无法清晰辨明，而这种模糊关系主要通过时间或空间两个维度来实现。

假如说融合自然的设计取态还是倾向于把建筑看成独立于环境的物体，图底关系模糊则追求建筑本质及内在结构上的突破。通过打破内外之间的界限，使建筑不再是独立的物体或空间，而是与环境渗透、不分彼此的“场域”，从而在场地内塑造一种新的秩序。这种建筑形态之所以出现，有赖于当代建筑学理论的快速发展与演变；例如斯坦•艾伦的场域理论就为这种概念模型的实现提供有力的理论依据。基于场域、混沌等新的视角，建筑不再是传统意义上的功能或历史象征主义的表达，而是趋向被认为是生物化、生态化、信息或能量流的一种表现。而数字化技术的高度发展也为这类空间形态的表达与实现提供有力支持。

扎哈•哈迪（Zaha Hadid）设计的伊斯兰艺术博物馆是其中一个体现图底关系模糊的例子。扎哈从卡塔尔的沙漠风貌吸取灵感，塑造宛如沙丘的波浪形曲线与色彩，形成与城市肌理一体化的景观。博物馆平面是由一组自西北向东南延伸的曲面组成，入口设在北面。起伏的屋顶造

2.3 Ambiguous Figure-ground Relationship

Normally, the figure-ground principle is used to describe the shape and proportional relationship between buildings and urban spaces. Nowadays, it has been considered as a way of thinking and a design logic to achieve N-architecture. Through blurring the figure-ground relationship, architecture obtains a new meaning in its surroundings, resulting in the obscurity between interior and external spaces. Generally, this ambiguous relationship is established by means of space and time and their resulting continuous movement.

In comparison with other conceptual models of melting with nature, an ambiguous figure-ground relationship tends to achieve structural and inherent integration between the two elements. Through breaking the boundaries between inside and outside, architecture no longer stands alone, but as a field condition merging with the landscape. Indeed, the development of contemporary architectural theories, such as Stan Allen's field theory, have provided a strong theoretical basis for the achievement of this design approach. Based on the new perspectives, including chaos and complexity science and field theory, architecture no longer merely expresses functional and historical symbolism as in its traditional sense, but is considered a biological, ecological, information entity or even a manifestation of energy flows. Incidentally, the rapid development of digital technology also has offered powerful

型，不仅把自然光引入建筑空间内部，也是对沙漠自然风貌的抽象化演绎。功能空间布置在曲面屋顶之间，与庭院及屋顶界面相互穿插，以此实现室内外空间界面的模糊和消解，人在步行体验中也因而形成错觉，产生室内与室外界限的不确定感，时而抽离，时而融入，并在此过程中同时经历自然与建筑环境的荣美。在这种空间模式中，自然景观成为空间组织的要素；或者说，空间被看成自然景观的一部分，建筑师通过建构两者的三维关系，来达到一体化的目的（图2-7a、b、c）。

tools for the representation and accomplishment of various morphological forms.

The Museum of Islamic Art proposed by Zaha Hadid can be viewed as a good example embodying the concept of an ambiguous figure-ground relationship within a unique urban environment. In this project, Zaha drew most of her inspiration from the landscape of Qatar. As a result, the whole structure took its form from the curves and colours of sand dunes and the movement of waves, immersing itself into the surrounding fabric. The structure undulates from the north-west to the south-east corner of the site, with its main public access at the north end. The public lobby equipped with a terraced plateau has a good view over the harbour, playing a major role in connecting all the public spaces.

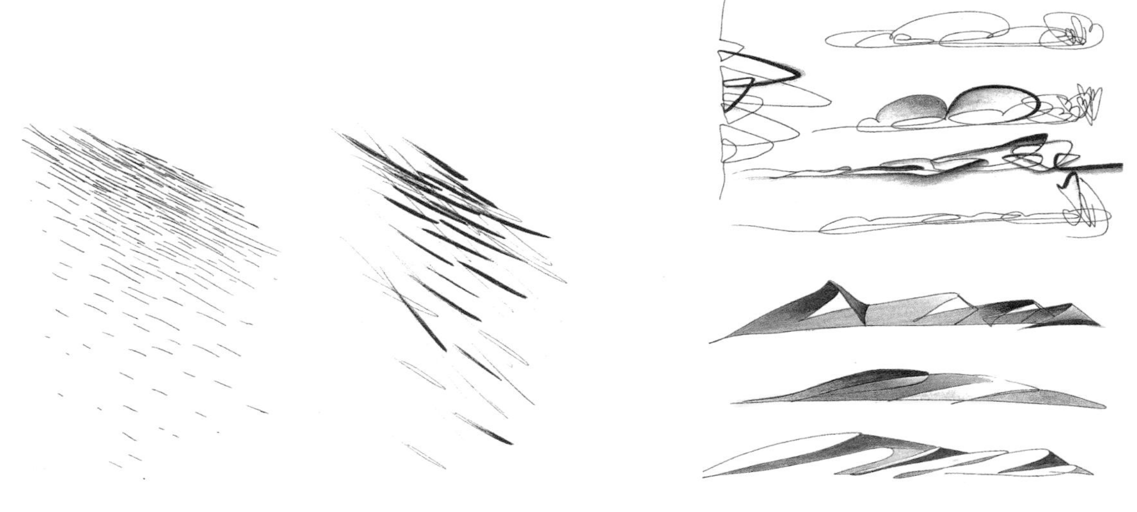

图2-7a 扎哈•哈迪的伊斯兰艺术博物馆方案概念草图

Programmatic spaces are arranged under the curved roofs, and the landscape, as an important element, participates in the form generation. Natural light is introduced into the building through the deep slits slashed in the undulating roof, whose surface is reflective in nature so as to reduce heat gain. When experiencing the spaces inside the museum, visitors are easily to be stimulated by the sense of uncertainty of indoor and outdoor boundaries. They feel involved sometimes, but detached at other times. Resulting from an experience like this, visitors can experience the beauty of nature and architecture simultaneously.

In this conceptual model, it is noticeable that the natural landscape functions as an essential element in the process of spatial composition. Likewise, architectural spaces are also regarded as parts of the natural landscape. Through establishing a three-dimensional interface between the two, the sense of N-architecture is formulated.

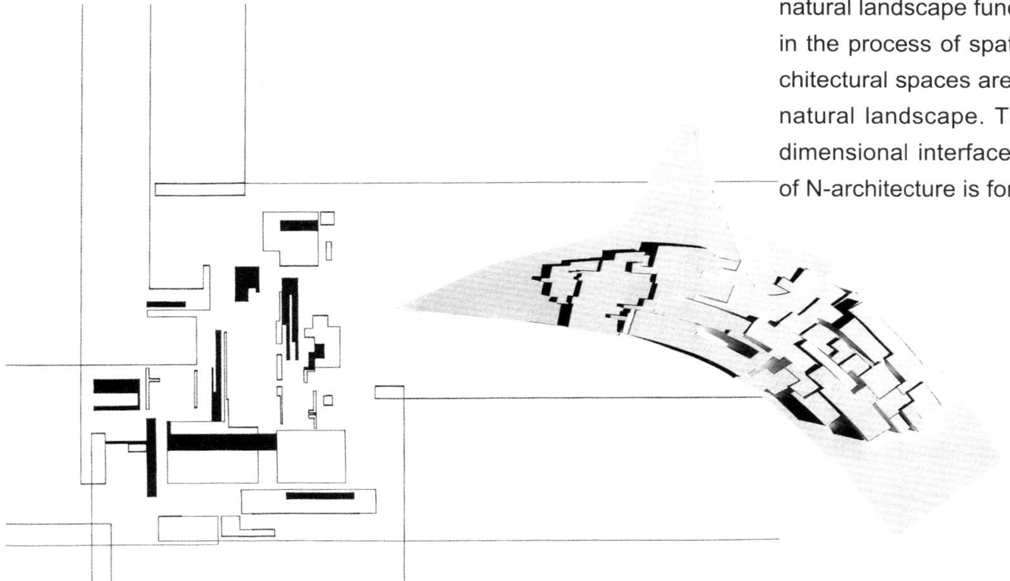

图2-7b　扎哈·哈迪的伊斯兰艺术博物馆方案概念模型

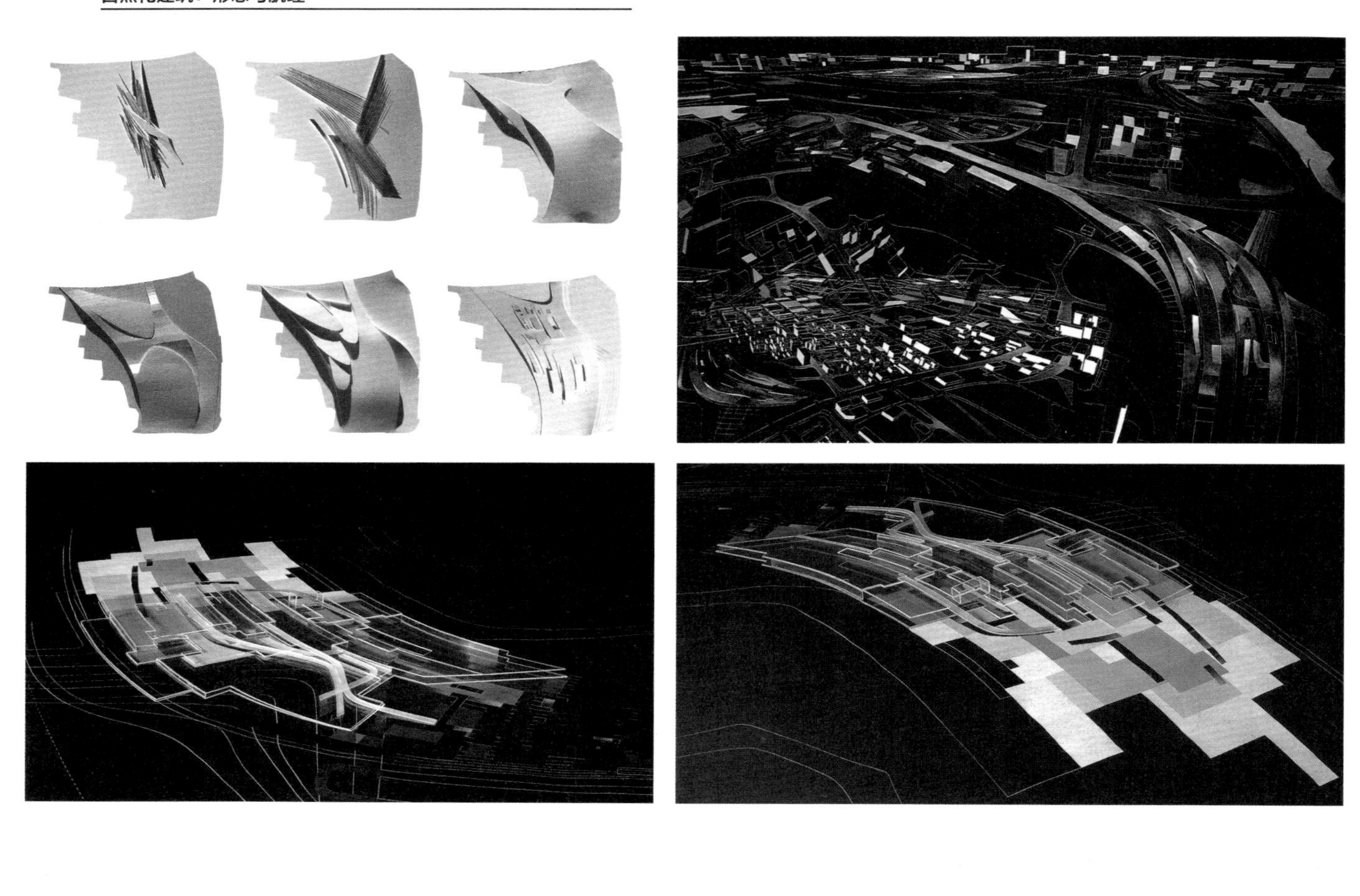

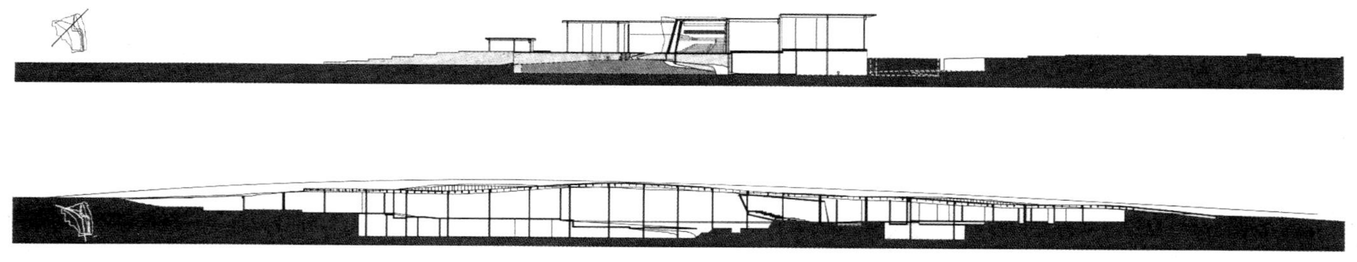

图2-7c　扎哈·哈迪的伊斯兰艺术博物馆方案

2.4 折叠与分支

传统建筑学常忽略空间的事件性（Events），忽略建筑的动态性及对人的行为模式的影响作用。建筑环境常被假定由两个处于静止状态的部分组成：建筑与环境（也就是图与底的关系）。为突破这种界限，当代许多建筑师，如彼得•艾森曼（Peter Eisenmen）、格雷•林恩（Grey Lynn）等，运用折叠理论进行新的设计探索，通过连续及多层次的连接方式把建筑与自然环境，或建筑内部的不同空间要素连接成整体。折叠建筑中，建筑界面被看作是连续的，在垂直与水平两个向度模糊或对图与底的关系进行重构，从而打破笛卡尔定义的空间秩序。折叠理论出现于 20 世纪末，引用林恩的定义是：“折叠就是把一些独立不相关的元素或事物同化成恒定的混合体。”艾森曼作为折叠建筑的创立者之一，也强调“weak form”，意即建筑的形态要以灵活多变的形式取得与环境的和谐。折叠建筑发展至今，随着数字化技术的成熟，其表达与实践也呈现多元并存的格局（图 2–8）。

2.4 Folding and Bifurcation

As mentioned, traditional theories often overlooked the idea of events occurring in architecture, as well as ignoring the influence of dynamic spaces on the pattern of human behaviour. Rather, it assumed that there existed two static conditions in the built environment: figure and ground. To break the tradition in design fields, many contemporary architects, such as Peter Eisenmen and Grey Lynn, have worked with the concept of folding in their design works. Through creating continuous interfaces and multi-level connections among different elements, they have produced a series of seamless spaces and forms.

Folding theory appeared in the late twentieth century. According to Grey Lynn's definition, folding means assimilating independent or unrelated matters into a constant mixture. In light of the ideas, architectural elements are coherent so that there

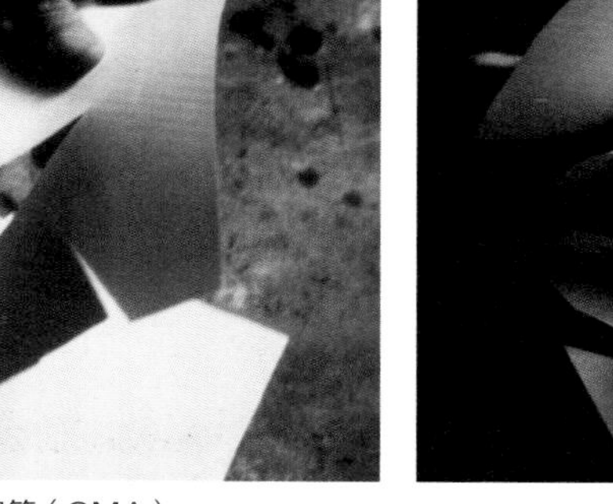

图2–8　折叠建筑（OMA）

折叠与分支在当代已成为创造形式与空间的重要手段，并被世界各地建筑师广泛采用，以解决各种设计问题。其优势在于，当各种元素需要在空间上实现多重和无缝连接，并形成一体时，折叠与分支就提供一种实现连接和组织空间的方式。较早运用折叠与分支理论进行设计创作的有 FOA（Foreign Office Architect），他们设计的横滨国际码头港就因成功演绎折叠建筑形态而获得公众的赞赏，并成功地为他们赢得国际上的声誉。码头港流畅的空间形态不仅引起人们对海湾波浪的浮想，同时也为公众提供一个翻卷起伏的滨水开放空间。码头港是一个长约 600 米的钢结构体系，始于一个入口广场，慢慢延伸形成商店、餐厅、等候等空间。设计概念源于创造一个介乎海港与花园之间的媒介体的构想，因此通过对建筑界面的连续性折叠，码头港实现不同空间层次的连接，并使空间流动性明显加强，从而使人与自然，建筑空间与空间之间的关系彻底融为一体（图 2-9）。

are no boundaries between the vertical and the horizontal, between the figure and the ground, breaking up the existing Cartesian order of space. In addition, being one of the founders of folding architecture, Eisenmen stresses 'weak form' in his practice, implying that architecture should embody an organic form in order to be in harmony with its natural environment.

Nowadays, with the development of digital technology, the practice of folding architecture has increasingly demonstrated a multi-variate pattern. Folding and bifurcation, an effective formal operation, have been widely used in numerous projects. It is noticeable that this tactic is not just a theoretical device. On the programmatic level, folding and bifurcation can be applied to solve particular design problems, notably when disparate elements of a site (or a building) need to be related in such a way that they appear seamless. More specifically, folding and bifurcation are effective in solving the problem of junction and composition.

One of the most famous examples of folding architecture in the world is the Yokohama International Terminal designed by FOA (Foreign Office Architect). In this project, FOA successfully interpreted the folding configuration of form and space, and finally built up their reputation in the field.

The terminal has not only evoked people's imagination of waves of the bays where it is located, but also created a naturally rolling open space for the public. The 600-metre steel structure, consisting of a plaza leading to the terminal, accommodates various facilities, including shops, restaurants, ticketing and waiting areas. Through creating a mediation between the harbour and the garden, the terminal has reacted against the rigid segmentation usually found in mechanisms dedicated to maintaining borders. Through continuous folding and bifurcation, the whole structure has created such a condition that the boundaries between different states are blurred, and various facilities are articulated throughout the constantly changing form. As a result, users cannot identify the structural elements from the envelope, and integration among people, architecture and nature has been successfully achieved.

图2-9　横滨国际码头港（FOA）

2.5 地景建筑

另外值得一提的是当代新兴的一种重要建筑理念：地景建筑。广义上的地景建筑与自然化建筑在概念上存在交叠，都是泛指所有亲近自然场地、体现建筑的自然特性、讲究与当地文化和人文关系契合的建筑理念。而狭义上的地景建筑，指的是和地景（地形、地貌、土地肌理等）高度融合的一种建筑形式，常融合模拟、隐喻、覆土、甚至折叠等多种手法而创作出来的一种空间形态，是当代建筑多元化创新的重要途径之一。

查尔斯•詹克斯的“宇宙秘密”可以说是一个早期的地景设计，其旨在创造一个具震撼力的象征性景观建筑（图2-10）。在这件作品中，一部混凝土楼梯如瀑布一般，蜿蜒而下，其步级最终消失在黑暗的水中，代表了宇宙起源的奥秘。另一个代表性地景设计作品是位于西班牙加里西亚，由彼得•艾森曼设计的文化中心；在这个设计当中，艾森曼通过一系列的建筑策略成功地塑造类似延绵不断山丘的风貌（图2-11）。而另一当代著名建筑师BIG，也因创作了一系列的地景建筑而蜚声国际；其代表作品包括爱彼得螺旋形博物馆（图2-12）和丹麦国家海事博物馆等。

2.5 Landform Architecture

It is worth mentioning that landform architecture is another influential conceptual model. In a broader sense, its definition has some overlapping with that of N-architecture discussed in this book. That is, both of them present the concept of nature, and express the natural characteristics of architecture by means of particular spatial patterns and forms. Also, they all pay sufficient attention to local culture, human society and local climatic conditions. However, landform architecture more specifically refers to structures that possess a high degree of integration with the natural terrain, such as topography, land texture, and so on. In most circumstances, its spatial form derives from a combination of metaphor, symbolism, melting, and even folding tactics, and currently it is treated as one of the pivotal measures to produce diversified and innovative contemporary architecture.

Charles Jencks' Universe Cascade can be considered as an early landform design for the sake of creating a strong symbolic landscape. In this work, a concrete staircase stretches like a waterfall, with its steps finally disappearing into the dark water below, representing the mystery of the origin of the universe. Another representative building is the City of Culture of Galicia designed by Peter Eisenmen in Spain. In this project, Eisenmen has succeeded in creating a sort of landform simulating a rolling hill by means of architectural measures. Another contemporary architect, BIG also has established their reputation in creating a series of landform architecture, including the Spiral-shaped Museum of Audemars Piguet and Danish National Maritime Museum.

图2-10　查尔斯·詹克斯的“宇宙秘密”

图2-11　彼得·艾森曼的西班牙加里西亚文化中心

图2-12　BIG的爱彼得螺旋形博物馆

2.6　衍生式设计

有别于其他自然化建筑形态，衍生式设计无论在理念还是在手段上都带有一定的颠覆性。它是借助于计算机技术，基于一组规则或编码来生成形体或空间的。目前，大多数衍生设计基于参数化建模，是一种对各种设计可能性进行快速探索的有效工具，目前已广泛地运用于包括艺术、建筑学、产品设计在内的设计领域。而在建筑学领域，它主要用于探索形式创作及结构性模拟。衍生式设计倾向于把建筑看成一个生物系统，或具有类似生物的属性。因此，此类设计常从学习自然界的内在规律出发，模拟自然界生物的行为模式；模拟的对象可涉及自然界各种尺度的事物，从生物体、器官，到细胞、分子结构等，以此来创造新的形式语言或建立新的空间环境。衍生式设计由于数字化技术的日新月异而迅速崛起，并已成为建筑学科的独立分支。各种编程环境（如 Processing）及软件（如 Rhino）的涌现，也使衍生设计得以广泛推广与运用。目前，衍生设计无论在建筑学教育还是在先锋性的建筑实践都占有一席之地，下面的案例介绍了衍生式设计的过程与手段，展现了自然界的现象或条件如何被转化为一组编码与参数，然后通过数字化技术，生成复杂多样的空间与形态。

“RAICES”（西班牙语“根”的意思）的设计作品是笔者与来自其他国家的三名设计师在英国建筑联盟学院（AA）的胡克公园校区（Hooke Park）共同设计与建造的一个凉亭建筑（pavilion）。设计概念始于对所处的森林环境的思考；设计任意选取若干树木作为参考点，

2.6 Generative Design

In comparison with other N-architectural morphologies, generative design features a high level of innovation and a certain subversiveness in both thinking and techniques. By taking advantage of computer technologies, it generates images, sounds, models, and animation based on a set of rules or an algorithm. Through parametric modelling, generative design has been regarded as an effective tool to explore a variety of design possibilities rapidly. Currently, generative design has widely been adopted in design fields, including art, architecture, communication design, and product design nowadays.

Generative design in architecture (also referred to as computational design) is mainly applied for form-finding processes and for the simulation of architectural structures. It is inclined to treat architecture as a biological system, having biologically similar properties. With reference to the inherent law or the process of natural lives, this method can simulate biological behavioural patterns at different scales, from an organism, organ, a cell or molecular structure, creating a new design language or a new spatial environment. In the architectural domain, benefiting from the rapid development of digital technology, generative design has increasingly turned into an independent discipline, and has gained its ground in architectural education and in design practice. Largely due to the new programming environments (Processing, Quartz Composer,

界定凉亭的边界并进行连接，然后通过Processing平台进行Bundling运算以模拟树根从地里延伸至地面的场景与动感，并结合空间高度及其他设计参数的设定，进行形态上和结构性的优化，最后确定凉亭的三维形态模型。这个设计向我们展示，设计从一开始至最后阶段都基于参数化的演算逻辑，避免人工化的干预与操作；由于设计结果基于某种设定的密码而自然衍生，建筑获得了如同自然界生命体般的特性（图2-13~图2-14）。衍生式设计开拓了设计思想的另一个领域，为形式创造提供更多的可能性。世界许多著名建筑师如扎哈·哈迪的建筑师事务所，就一直致力于这个领域的研究与实践。

Open Frameworks) or scripting capabilities (Rhinoceros 3D, Scriptographer), generative design has become easily operated even for designers with little programming experience to implement their ideas.

The following example illustrates a typical process of generative design, where certain natural phenomena or conditions were translated into a set of parameters in order to produce complex and diverse spaces and forms.

The pavilion named 'RAICES' ('root' in Spanish) was designed and fabricated by the author and three other international designers in the Hooke Park Campus of AA School. The design concept is

图2-13a　凉亭建筑："根"

initiated from the thinking of the forest environment where the pavilion is located. By selecting a number of trees as reference points arbitrarily, the design team first set the boundary of the pavilion, subsequently connecting them into a network. And then, a bundling operation was carried out through the Processing Program to simulate the roots growing out of the ground, forming the model of the pavilion. Meanwhile, by taking into account other design parameters, such as spatial and programmatic requirements, the pavilion model was optimized both morphologically and structurally. From this design, it can be implied that generative design, from its beginning to the final stage, is based on a set of parameters resulting from contextual analysis or specific design ideas rather than by manual operation. As a result, the architecture acquires the likeness of natural life, growing and evolving automatically based on certain codes as set. In this sense, it comes as no surprise that generative design has been treated as a powerful way to explore various design possibilities. Many world-famous architects, such as Zaha Hadid and her design team, have been committed to the extensive research and fabrication in this field.

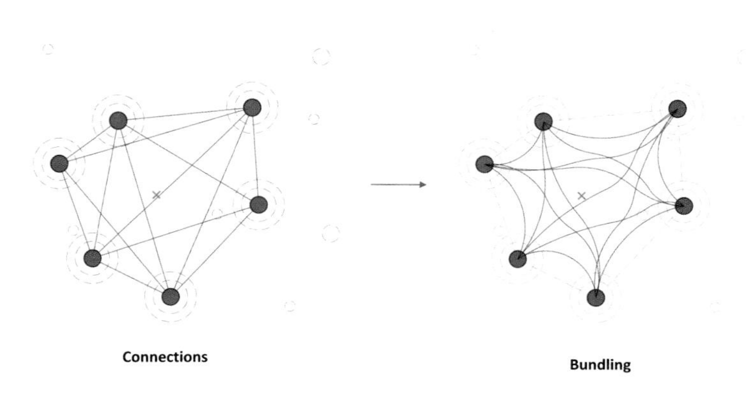

图2-13b　选取六棵树定位并进行Bundling 运算

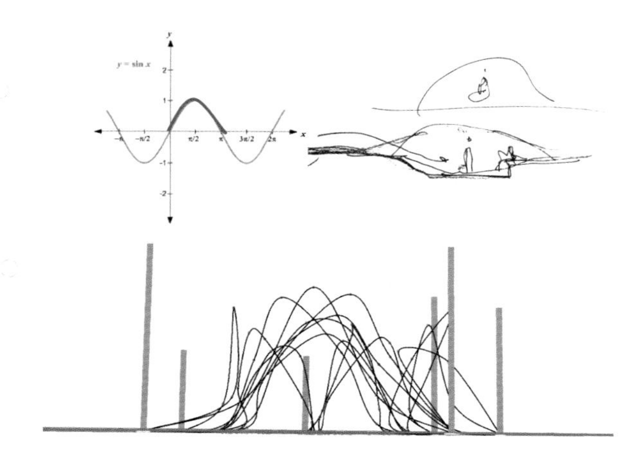

图2-13c　以使用高度选取空间尺度

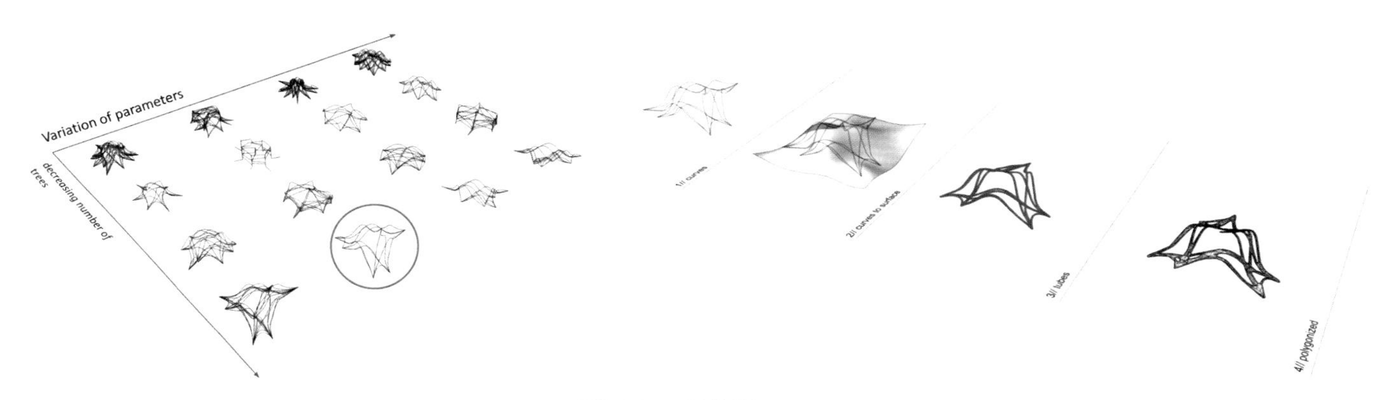

图2-14a　形态推敲与选择

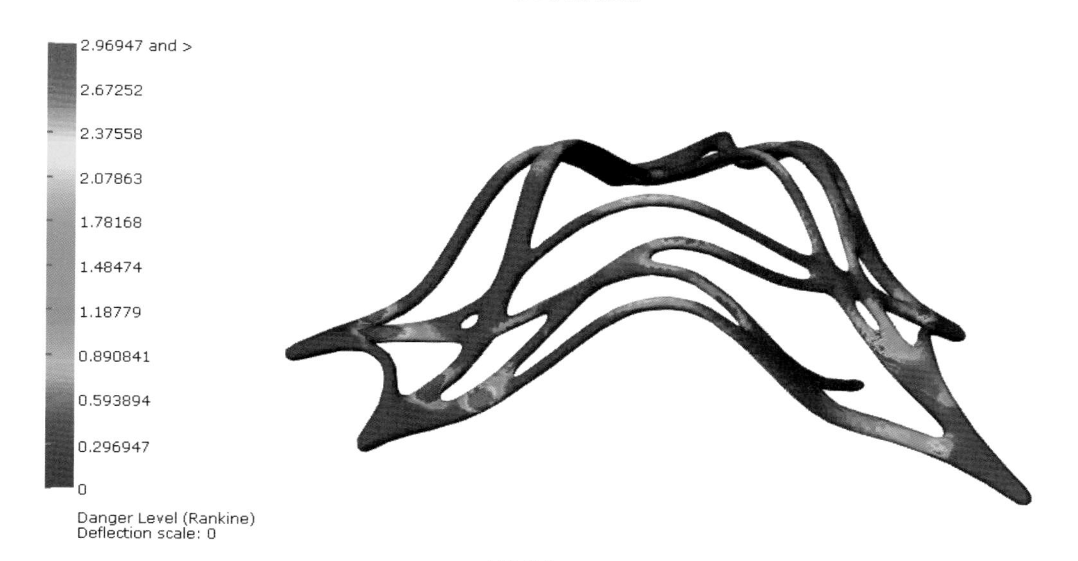

结构优化

图2-14b　“根”的生成设计过程

2.7 可持续发展理念下的生态设计

自20世纪70年代开始，伴随着可持续发展的理念，生态设计就日益成为建筑学领域的重要课题；而在生态建筑研究以及实践当中都有突出成就的代表人物是马来西亚建筑师杨经文。杨经文曾在英国建筑联盟学院（AA）就读，并在英国剑桥大学（University of Cambridge）取得博士学位。其回到马来西亚执业后，便致力于生物气候建筑的研究和实践。其早年成立的 Hamzah & Yeang 事务所就基于生态感应原理进行相当数量的摩天大楼的设计，并发展出以被动式、低能耗为特征的“生物气候设计”方式。杨经文认为，真正的生态设计应该是建筑环境与自然环境的完美和良性的融合：是系统的、物理化的、可实施的。

生态设计并不是一个单纯在技术层面来实现的问题，还涉及社会、文化及经济等向度的生态发展，同时还关注建筑的地域性。杨经文认为，所谓地域性建筑应不懈追求无形的地方场所精神，反映当地的乡土及文脉特征，并与当地的实际需求以及人的情感寄托产生密切的联系，从而使建筑能在特定的地区通过其造型美学、组织构造、技术整体及材料来充分体现这些区域的内涵。另外，地域性建筑还应反映出当地特有的自然型态、社会场景、经济以及政治状况；而自然型态就包括了当地的气候、地形及生态条件等。一个真正意义的可持续性设计应综合考虑这些因素对建筑物可能造成的影响。基于这些理念，杨经文及其设计团队致力发展适应亚洲独特都市环境的建筑语言。例

2.7 Eco-design in Sustainable Development

Since the 1970s, along with the concept of sustainable development, eco-design has increasingly become an important subject in architectural fields. Malaysian architect Ken Yeang, who studied in AA School of Architecture and got his doctoral degree at the University of Cambridge, has attained outstanding achievements in both ecological research and practice. Since his return to Malaysia to establish his own career path, Ken Yeang has been constantly committed to the development of bioclimatic architecture, especially in tropical regions.

As a partner of Hamzah & Yeang Architect Firm at his early stage, Ken and his team focused more on the design of skyscrapers, subsequently developing the 'bio-climatic design method' featuring passive low-energy consumption. Ken Yeang believes that real eco-design should be able to achieve a perfect integration between the built environment and the natural landscape, and should be systematic, flexible and feasible as well.

Ecological design is not a simple question accomplished at the technical level only, but also relates well to the ecological development in social, cultural and economic dimensions, in addition to reflecting the particular characteristics of the region. In light of this concept, Ken Yeang believes that real local architecture should pursue the invisible spirit of the place, reflecting its local context and regional cha-

如，通过对太阳光的角度、风的动态、上升暖气流特性的详细分析，并结合当地的建筑美学，创作出大量具地方性特色的高层建筑。杨经文还认为，建筑物自身应是一个开放性的系统，能与环境相互渗透，这就像一个筛子一样，能够使外部气候与内部空间互相流通、交换，使内部空间达至一个最合适的状态；然而，更重要的是，建筑师对于当地环境、文化背景以及社会问题要有适当的考虑与回应。这些都是建筑师工作的职责,而这些努力将产生真正意义的地域性建筑。

杨经文设计理念在其设计的新加坡 EDITT 大厦有较完整的体现。虽然此项目最终并未建成实施，但却是杨经文第一次提出“线性垂直公园”及“垂直城市化”等生态设计概念的作品（图 2-15）。在此设计的基础上，杨经文又在其后来一系列作品中提出了诸如“生态细胞”“绿色生态基础设施”等颇具启发性与先锋性的概念，为生态设计的健康发展提供很好的思路。

racteristics, in addition to responding to the real situation and satisfying local people's emotional needs. In this way, architectural design is able to reflect specific meaning by taking into account unique local aesthetics, structural tectonics, technology and materials. Also, local architecture should reflect local natural patterns, social scenarios, economic and political conditions; and the natural pattern mainly includes local climate, topography and ecological conditions, and so on. Ken Yeang further pointed out that a real sustainable design should consider the impact of these factors on the performance of buildings. Based on these ideas, Ken Yeang and his team have produced a set of local design vocabularies and created a number of influential projects in high-density cities through detailed analyses of solar angle, wind dynamics, and air-flow characteristics, combined with the consideration of local architectural aesthetics.

In these projects, buildings are treated as open systems, able to be penetrated by the natural environment. Through achieving a climatic exchange with the external environment, the interior condition can be maintained at an appropriate standard. During their design process, sufficient attention is paid to the local environment, cultural and social issues, and it is believed that these efforts finally contribute to a regionalistic design in a real sense.

Yeang's design philosophy has been typically revealed in the design of EDITT Tower in Singapore.

Although the project was not implemented, it was the first time for Yeang to raise a series of ecological design concepts, such as 'vertical linear park' and 'vertical city', considerably contributing to the emergence of his other pioneering and provocative concepts, including 'eco-cell' and 'green infrastructure'. It is believed that Yeang's contribution to contemporary architecture is that his propositions and experience are a valuable reference for the worldwide development of eco-design in a positive way.

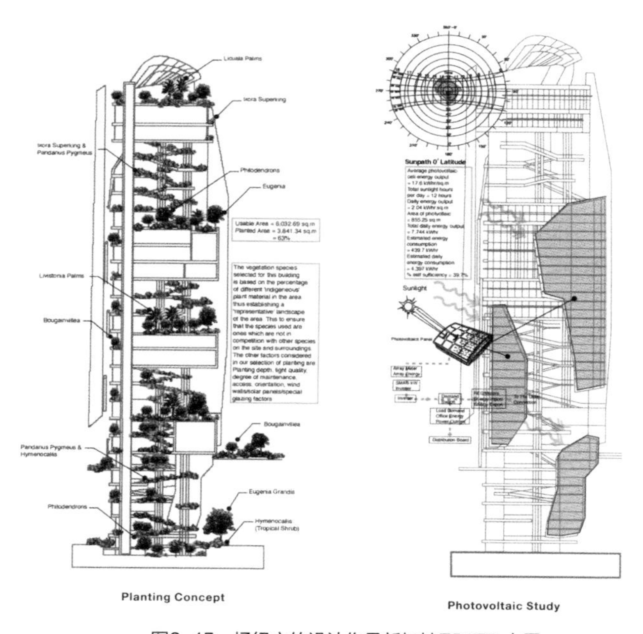

图2-15　杨经文的设计作品新加坡EDITT 大厦

2.8 小结

总而言之，当代建筑学科发展日趋多元并存格局，许多设计实践并不是基于单一概念模型来进行自然化建筑的探索，而是综合以上若干概念进行建筑创作的。毫无疑问，随着技术的发展和科学的进步，人们对自然的探索将越来越深入和广阔，而自然界的丰富与多元也必然为人类的创造性活动提供无穷无尽的源动力。在此背景下，我们相信，自然化建筑也将进一步演进和发展，并在未来产生更多的形态可能性。

2.8 Epilogue

It is worth noting that the development of contemporary architecture has increasingly displayed a multi-variate pattern, and a large number of design practices have been carried out based on the combination of various conceptual models. Attributed to the advancement of both science and technology, people are likely to explore nature more thoroughly than before. In turn, the richness and diversity of nature are bound to provide an endless source for human creativity. In this sense, N-architecture will inevitably be further developed and evolve, with more and more morphological possibilities coming into being in the coming days.

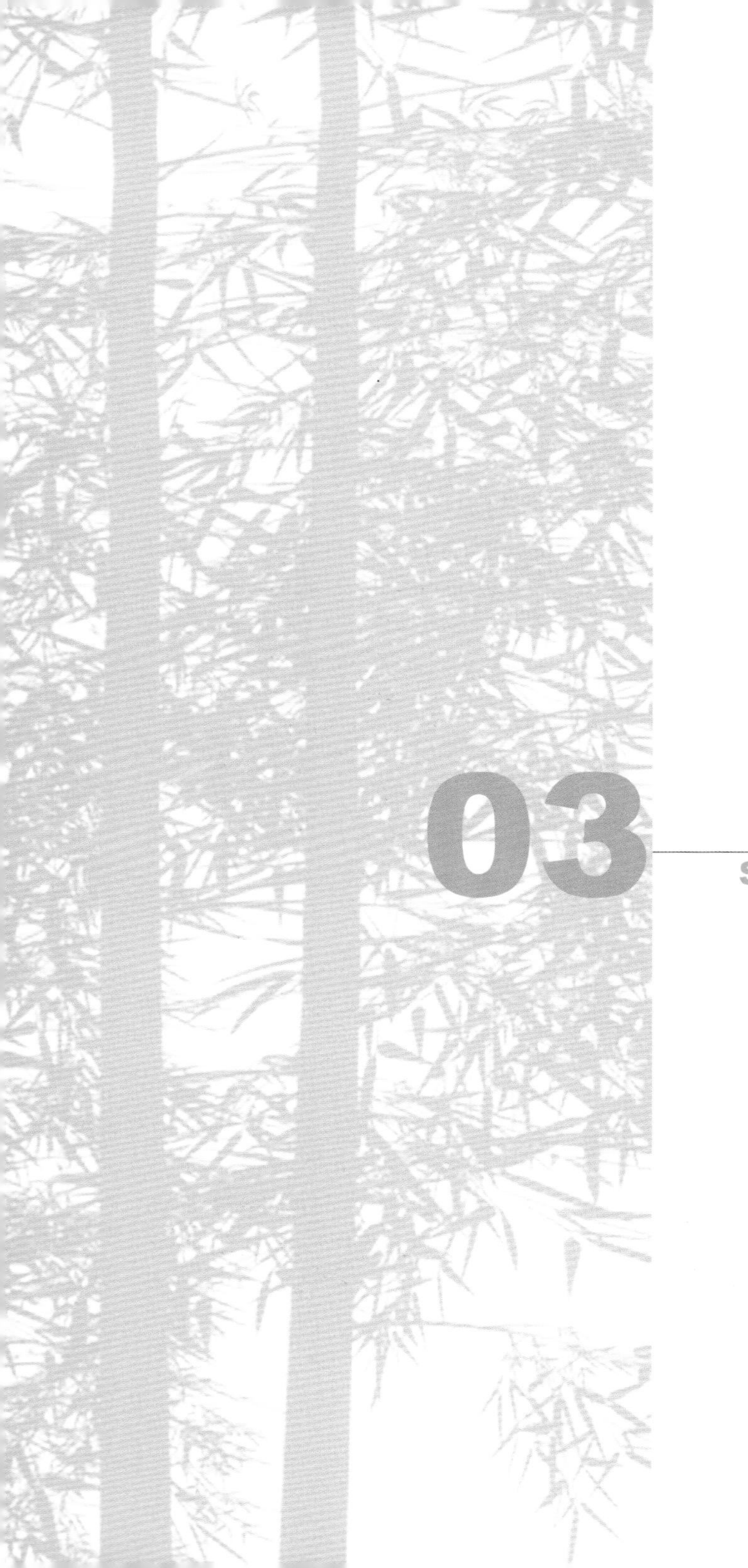

03 空间肌理分析：自然化建筑的设计实践

Spatial Pattern: Practice of N-architecture

这一章节通过对具体设计案例的介绍，描述一系列实现自然化建筑理念的具体策略，并展示其演化成独树一帜的建筑空间类型的思路和过程。

By introducing a number of design works, this chapter aims to demonstrate how the concept of naturalization is transformed into specific design strategies, further producing unique types of architectural space.

3.1 液态空间

液态空间是基于象征和模拟的手段而产生的一种空间类型。可以说，液态空间是基于非线性思维来解决特定建筑问题的。常态的设计观念倾向于把建筑看成一个独立的物体或静止的状态，而液态空间设计思想打破传统的界限，设想空间可以像液体一般，连续而流畅，以此来实现建筑与环境以及不同建筑空间要素之间的无缝连接，从而使建筑成为激发人的连续运动的引擎。液态空间所采用的策略包括空间螺旋化、交织、起伏和折叠等（图 3–1）。

3.1 Liquid Space

Based on the concept of metaphor or folding, one possible spatial type is liquid space, which is based on non-linear thinking to solve particular design problems. As mentioned, traditional design theories tend to treat architecture as isolated objects or static conditions. However, with the development of science and technology, architectural design has been allowed to go beyond its conventional boundary and create more possibilities. The rationale behind this spatial type is that architecture can be liquid-like so as to create a seamless relationship between architecture and nature, as well as between different parts of buildings. More than that, it also generates dynamic behavioural patterns inside the space.

螺旋

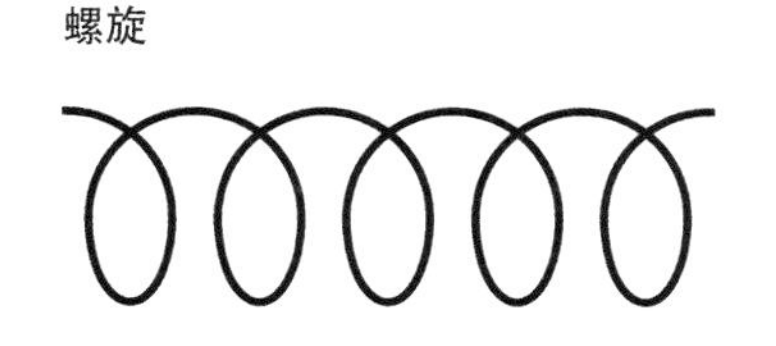

交织

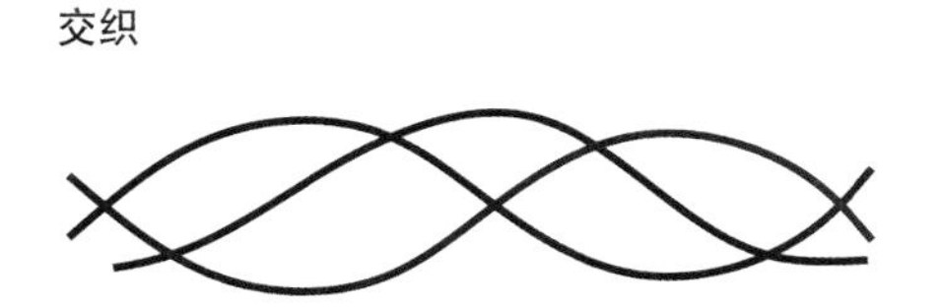

起伏与折叠

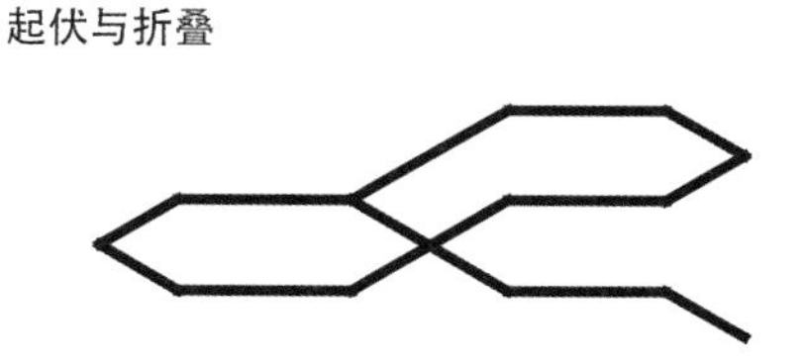

图3–1　液态空间的几种模式

3.1.1 “交织校园”：一个关于未来学校的设想

“交织校园”是一个关于未来校园设想的竞赛方案，并获一等奖。设计概念的产生源于对现有教育模式的批判性思考，因而提出由于信息科技的发展，学校在未来社会的功能与角色将发生演变的设想。在未来时代，现有的以教师为中心进行授课的教育模式将发生转变，逐步演化成以学生为导向的学习模式。基于此理念，传统的学校功能构成也将被打破、重构，取而代之的是由三条线性空间组成，代表了三种学习模式，分别为自主学习空间、指导与协作空间及自然绿化空间。自主学习空间是学生通过先进信息技术自主学习的场所，而指导与协作空间则是学生与教师或学生与学生之间互动的空间，而自然景观空间是一个心灵净化、创造力启示的场所。三类空间交织在一起，回应蜿蜒曲折的场地地势的同时，也如液体般流动穿插，从而衍生出未来校园生动的空间意象。在这里，各类型学习空间高度共生、杂交，但同时又结构清晰，正如未来的教育模式，多元复杂，同时又具备无限的灵活性与开放性。在此“交织校园”中，液态空间所激发的互动行为模式也将成为未来学校教育的有效组成部分（图3-2~图3-6）。

3.1.1 Campus in Weaving: A High School Design for the Future

'Campus in weaving' is a design work that won first prize in a future high school design competition held in 1999. Initiated from the rethinking of what the high school would be like in the future, the designer then proposed that with the development of information technology, students would increasingly be the core of education rather than the traditional one, where teachers generally dominate the learning process. Based on this, the programming of conventional school was broken and reconfigured. As a result, three linear spaces, including self-learning spaces, instructive and collaborative learning spaces, and a landscape lane were created respectively. Self-learning space provides students with the opportunity to absorb knowledge through advanced technologies, and instructive space is a place for interaction and collaboration between students and teachers. By contrast, the landscape lane is a dynamic place provided for spiritual purification and creativity cultivation. These three spaces are interwoven together, resulting in a vivid and liquid-like campus, well conforming to the topography of the site. Inside this campus, various types of learning spaces are highly integrated, yet distinct in structure, just like the education pattern in the future, complex and diverse but with infinite flexibility. Meanwhile, the continuous interaction and diverse activities conditioned by the campus would also become an integral part of the school's education.

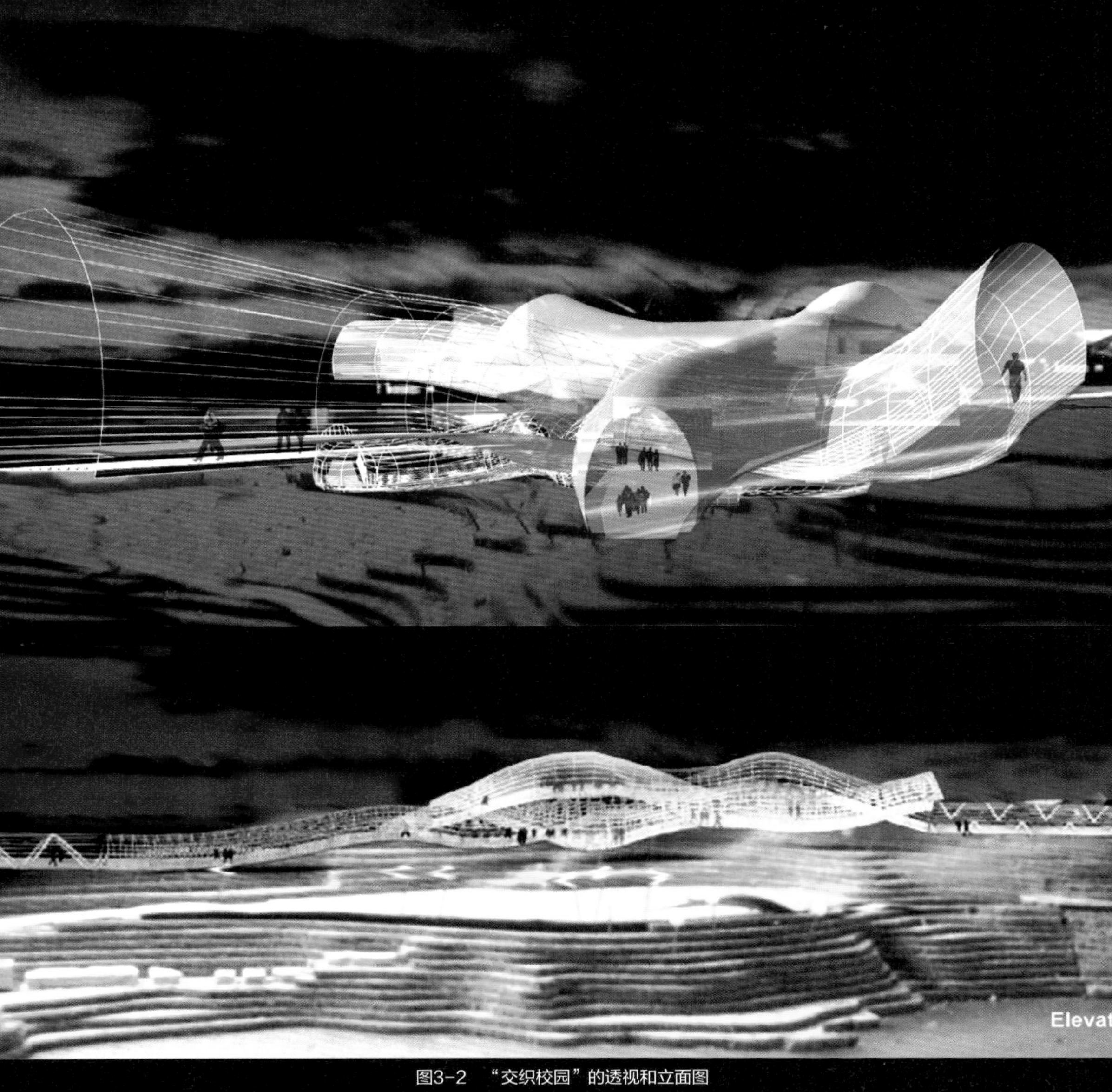

图3-2 "交织校园"的透视和立面图

图3-3　“交织校园”位于自然地形复杂独特的区域

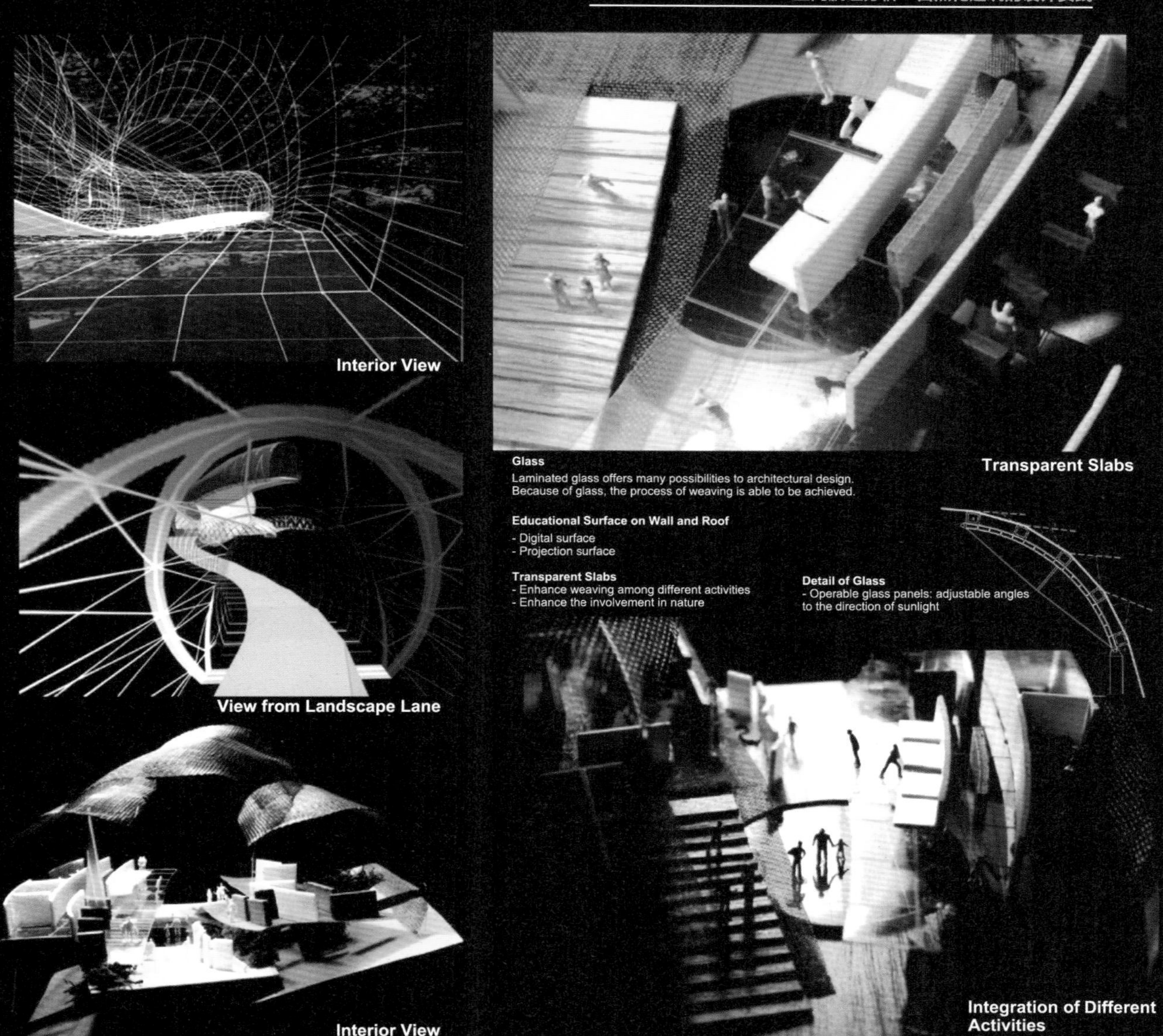

图3-4 “交织校园”节点空间

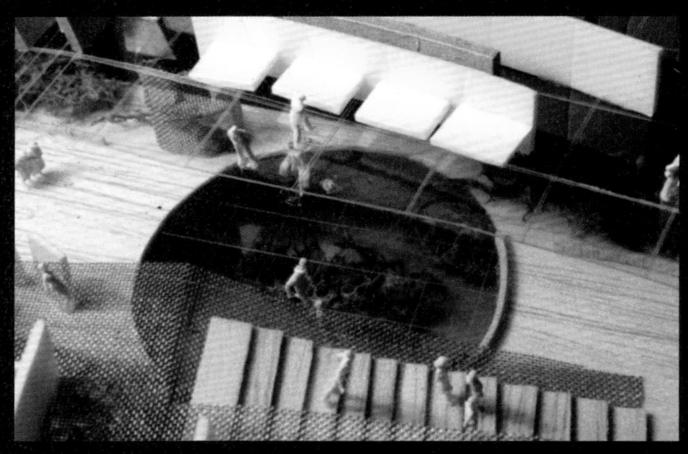

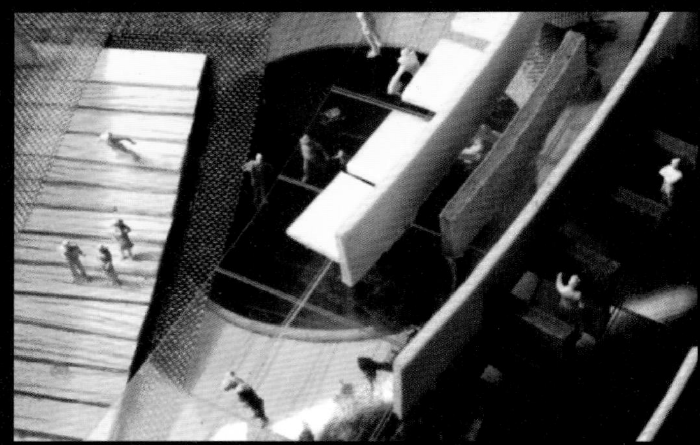

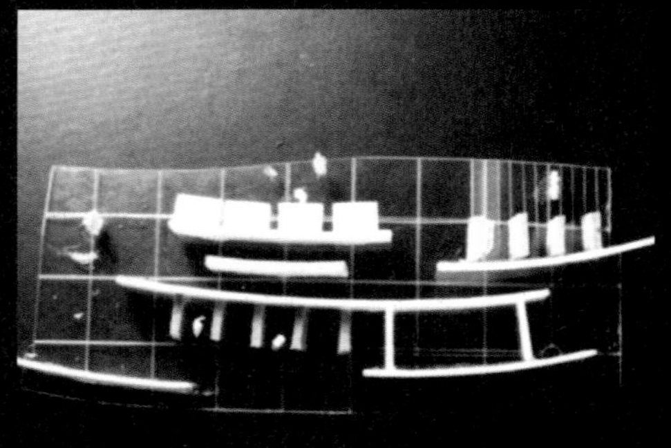

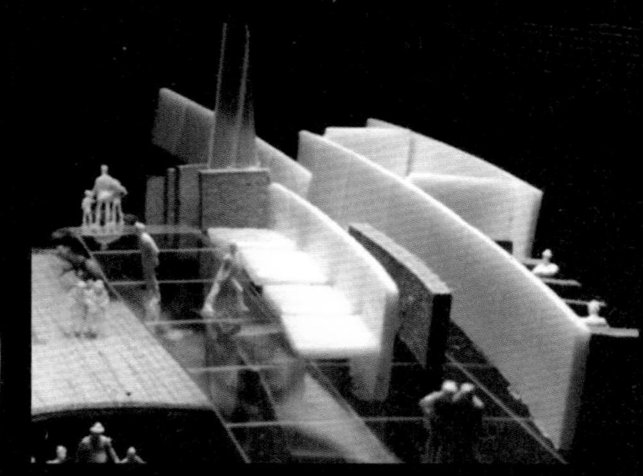

图3-5 “交织校园”设计模型

图3-6 “交织校园”设计模型

3.1.2 “水袖云舞”：岭南艺术博物馆

“水袖云舞”是一个关于岭南艺术博物馆的设计，其建筑形态也是液态空间类型的典型体现。基于对中国民间水袖舞的认识和理解，设计以自然化建筑为手段，模拟袖子舞动时的飘逸与动感，从而营造出一个连续而流动的空间体系，与蜿蜒曲折的湖岸相呼应的同时，也在观赏过程中激发人们与艺术品之间的互动与交流。在这种空间模式中，建筑与自然、建筑自身水平与垂直方向界面的关系已模糊。其结果是，人的行为活动有效地成为艺术展览的一部分，建筑整体很好地实现了人、艺术和自然的和谐统一（图 3-7~ 图 3-11）。

3.1.2 Sleeves in Weaving: Lingnan Art Museum

'Sleeves in weaving' is a design work of Lingnan Art Museum. As an awareness of Chinese folk sleeve dance, the designer created a liquid-like form by simulating the performance of sleeves. As a result, the liquid spaces generated by weaving not only function as an echo of the organic lake shore, but also encourage the constant interaction between visitors and art pieces. Inside the museum, the boundary between architecture and nature has been blurred, with the horizontal and vertical interfaces obscuring each other. As a result, visitors' behavioural patterns also become an integral part of the exhibition, substantially achieving a good harmony among people, arts and nature.

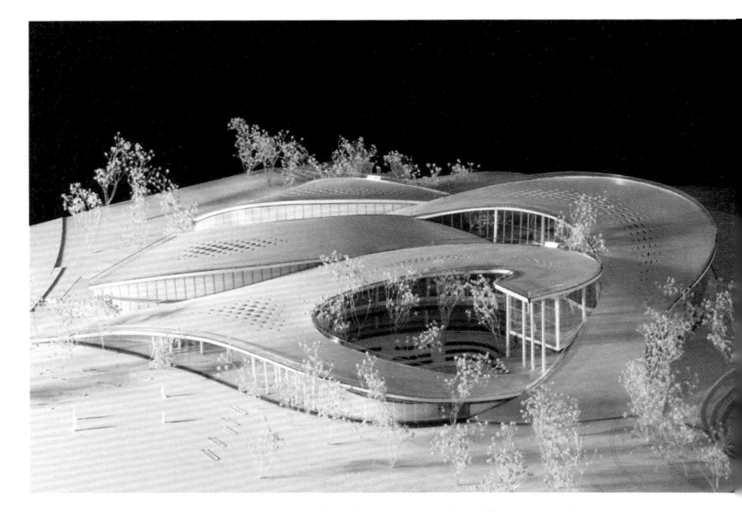

图3-7 “水袖云舞”艺术博物馆设计模型

ARCHITECTURE AS LANDSCAPE

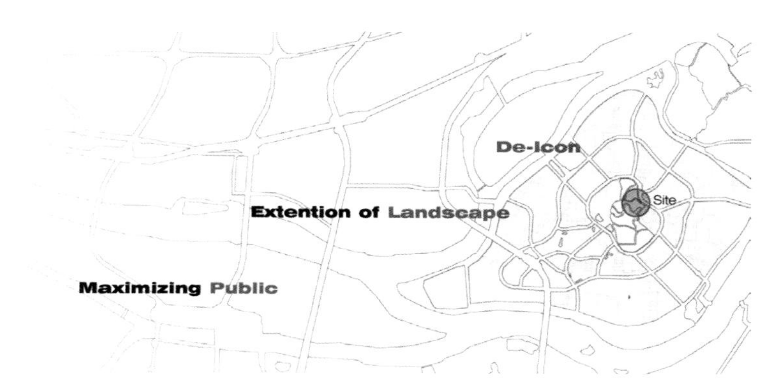

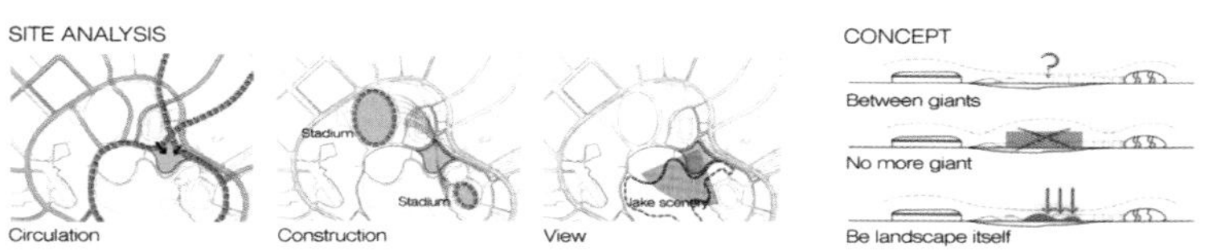

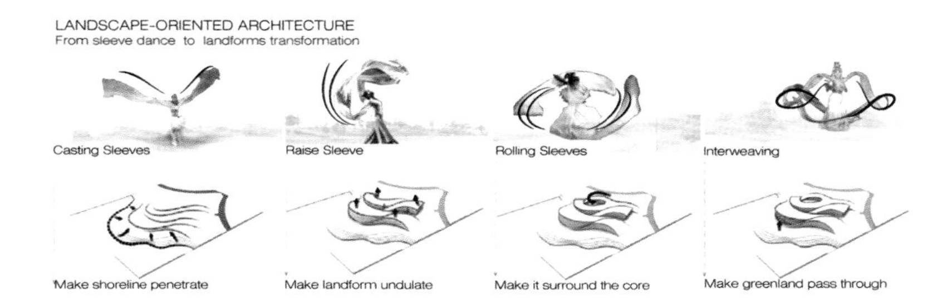

图3-8 “水袖云舞”艺术博物馆概念图

Air Meets Art

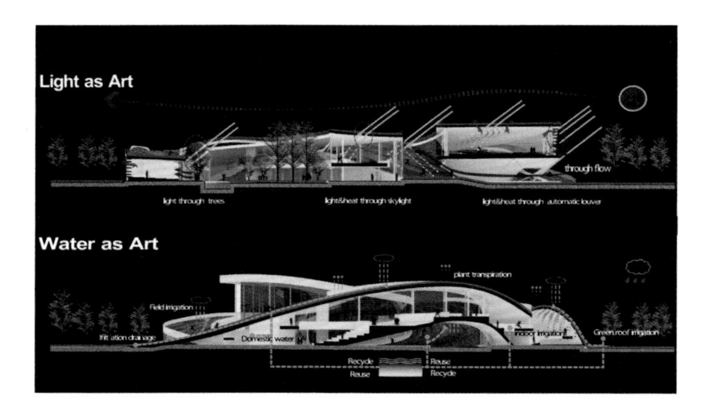

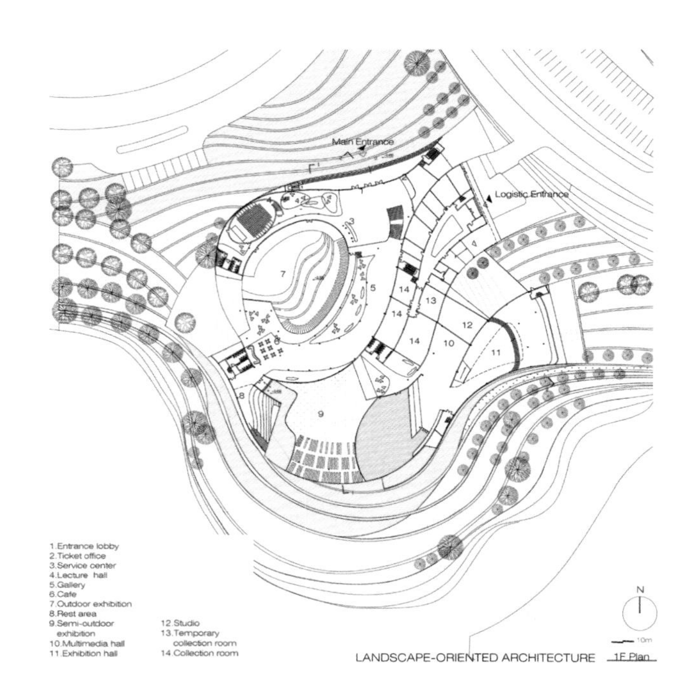

By interweaving linear land,
the Art Museum offers multiple types of space,
which leads to variety relationships between Art work and Nature.

Nature Interacts with Architecture

Art Interacts with Visitors

图3-9 “水袖云舞”的平面图及气候适应设计及空间分析

图3-10 “水袖云舞”设计模型

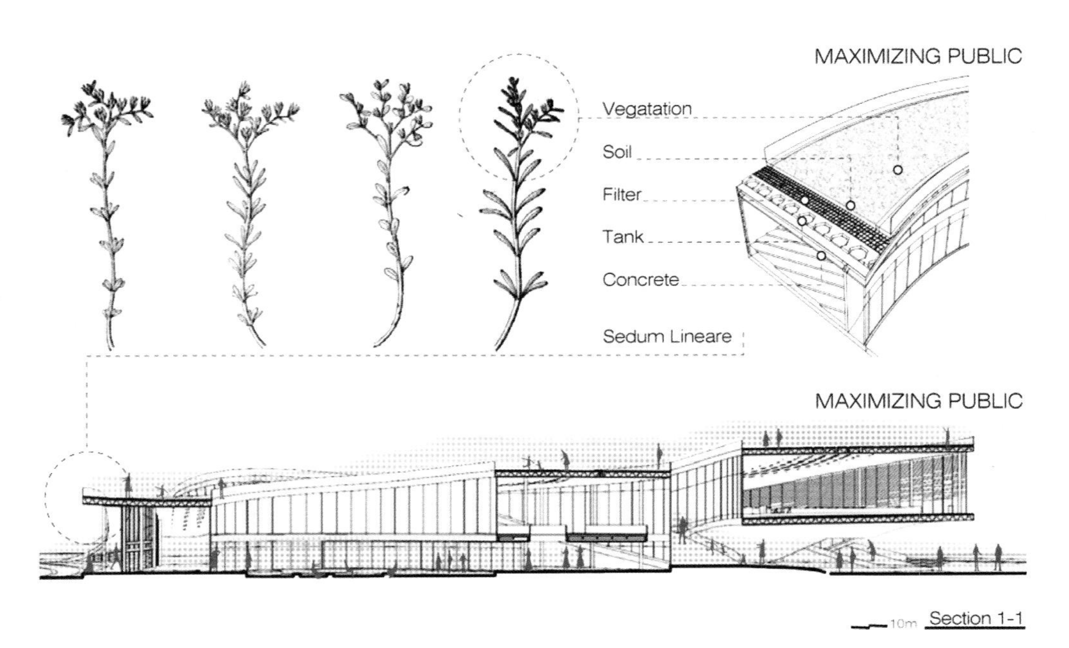

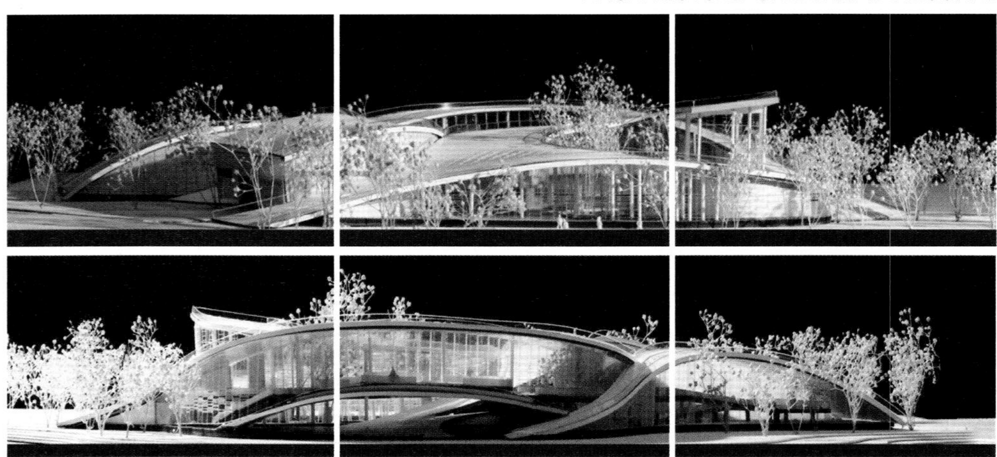

图3-11 “水袖云舞”设计模型

3.1.3 “网动艺海”：漳浦文化中心

“网动艺海”是一个文化综合体设计，包含博物馆、图书馆、档案馆等多种城市功能。场地位于中国福建省南部的漳浦县，是一个颇具历史文化底蕴的地区。为凸显漳浦海洋文化及城市快速发展的特点，设计取海浪及渔网的意境作为概念出发点。因此，建筑体量的生成始于一种液化的状态，慢慢演变成一种几何固态，象征着自然与人造环境的高度一体化。基于数字化技术的支持和演算，各建筑界面有效地实现无缝化，使整个建筑体量如同从海浪中衍生出来一般。在这种演变的过程中，液态空间也应运而生，在不同功能之间或者不同空间要素之间很好地实现了无缝连接，并为使用者提供一种独特的空间体验。得益于建筑形态的独特性，该文化中心建成后将成为一个地标，为市民提供文化交流及接受文化熏陶的场所。同时，该文化建筑群的建成也将成为一种驱动力，进一步促进城市的更新及发展（图3-12~图3-16）。

3.1.3 Moving-net: Zhangpu Cultural Centre

'Moving-net' is a cultural complex design, including museum, libraries, archives and other city functions. The site is located in Zhangpu County, the southern part of Fujian Province of China. This is a region famous for its geographic and cultural richness and diversity. In order to highlight the characteristics of local marine culture and the rapid development of the city, the architecture takes its physical form from the waves and fishing nets in order to express a condition, where the liquid state gradually evolves into a more regular geometrical form, symbolizing the high degree of integration between nature and the man-made environment. With the support and calculation of digital technology, various building interfaces finally have achieved seamless connections. As a result, the entire building form evokes an imagination as if the structure is derived from waves. Within the organic envelope, a system of liquid-like spaces also comes into being, achieving the continuity among various functions and further stimulating a dynamic movement pattern inside these spaces.

Owing to its unique architectural form, the cultural centre is deemed a landmark, providing citizens with a place to achieve cultural exchange and promotion. Meanwhile, the cultural centre is also regarded as a driving force, stimulating the urban development of the city.

图3-12 “网动艺海”文化中心鸟瞰

撒网的轻柔飘逸

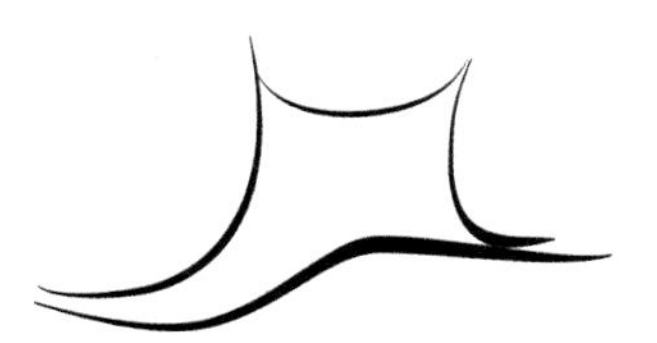

海浪翻卷的动感

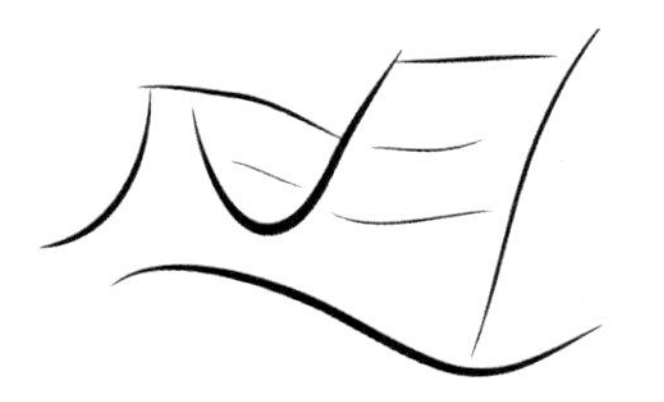

海浪冲刷岩石形成的流动界面
是对当地景观“十里画廊”的
抽象和解释

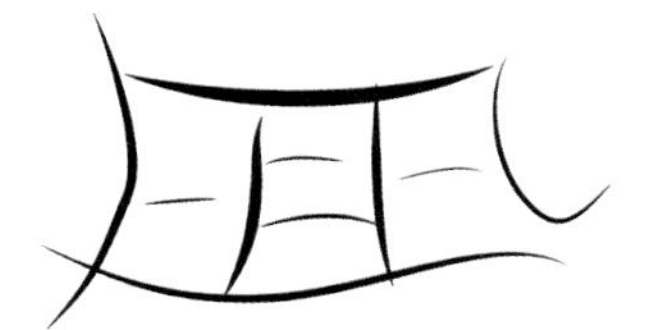

传统漳浦民居“凤尾”
屋顶天际线的启示

图3-13 “网动艺海”概念生成

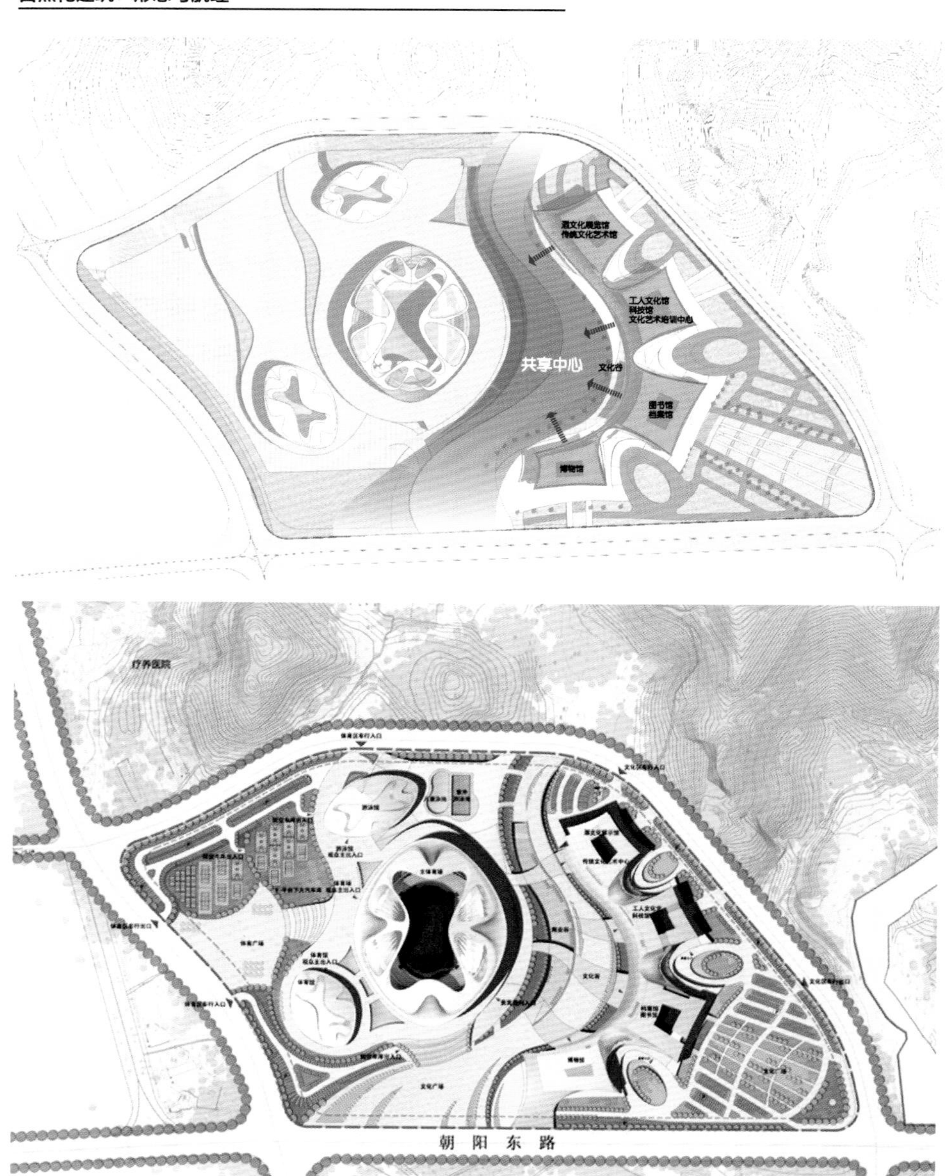

图3-14 “网动艺海”功能分区与总平面图

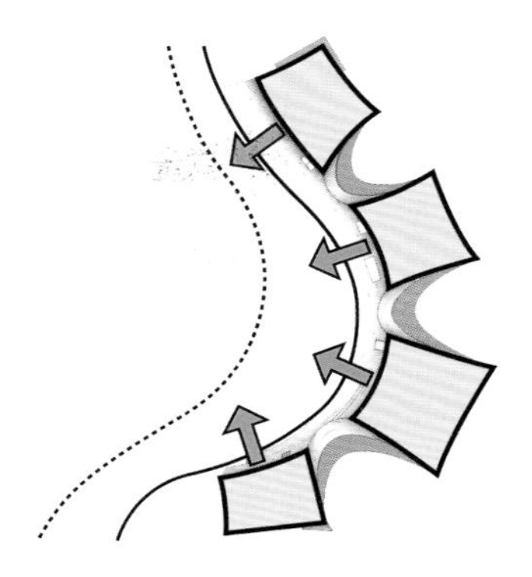

Step 1
呼应“文化谷”，建筑群体向心布置

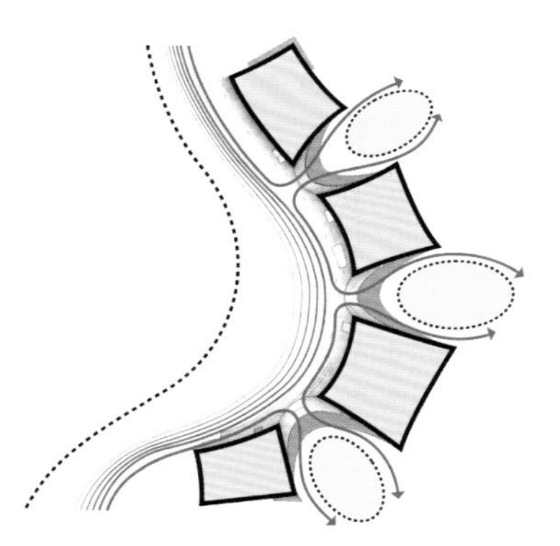

Step 2
“文化谷”景观向东面城市景观渗透，切割成四个多边形体量

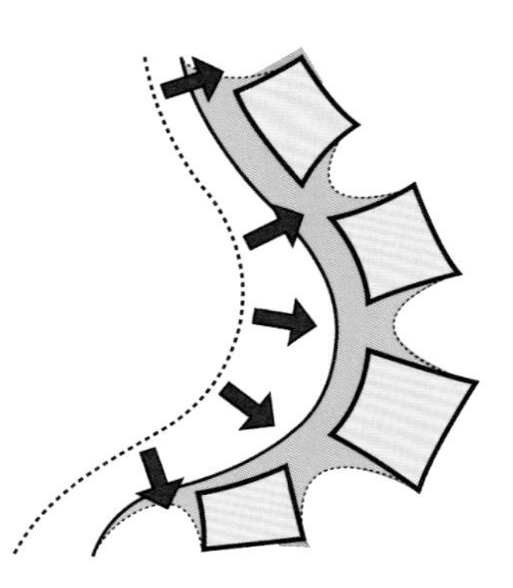

Step 3
谷的张力冲刷成四个具柔和界面的建筑形体

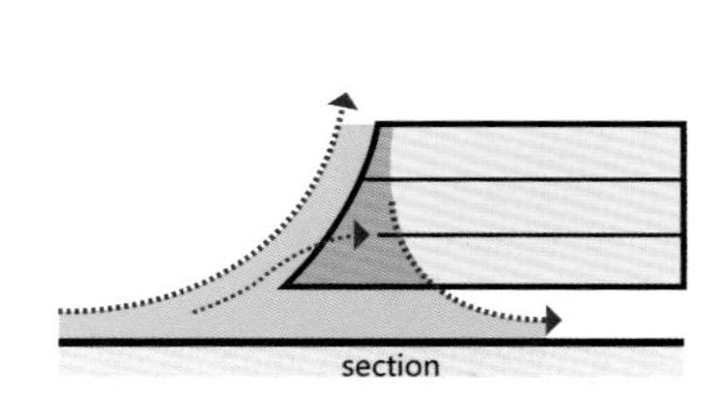

Step 4
软质界面竖向延伸

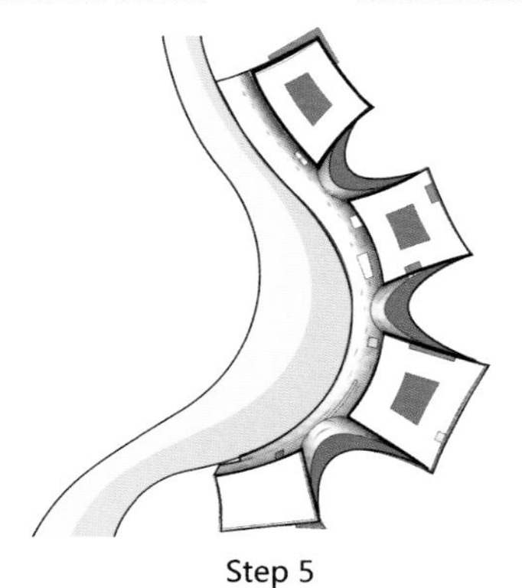

Step 5
形成动感延绵的建筑整体，表达自由形态向几何形态演变的一个过程

图3-15 “网动艺海”总体布局概念

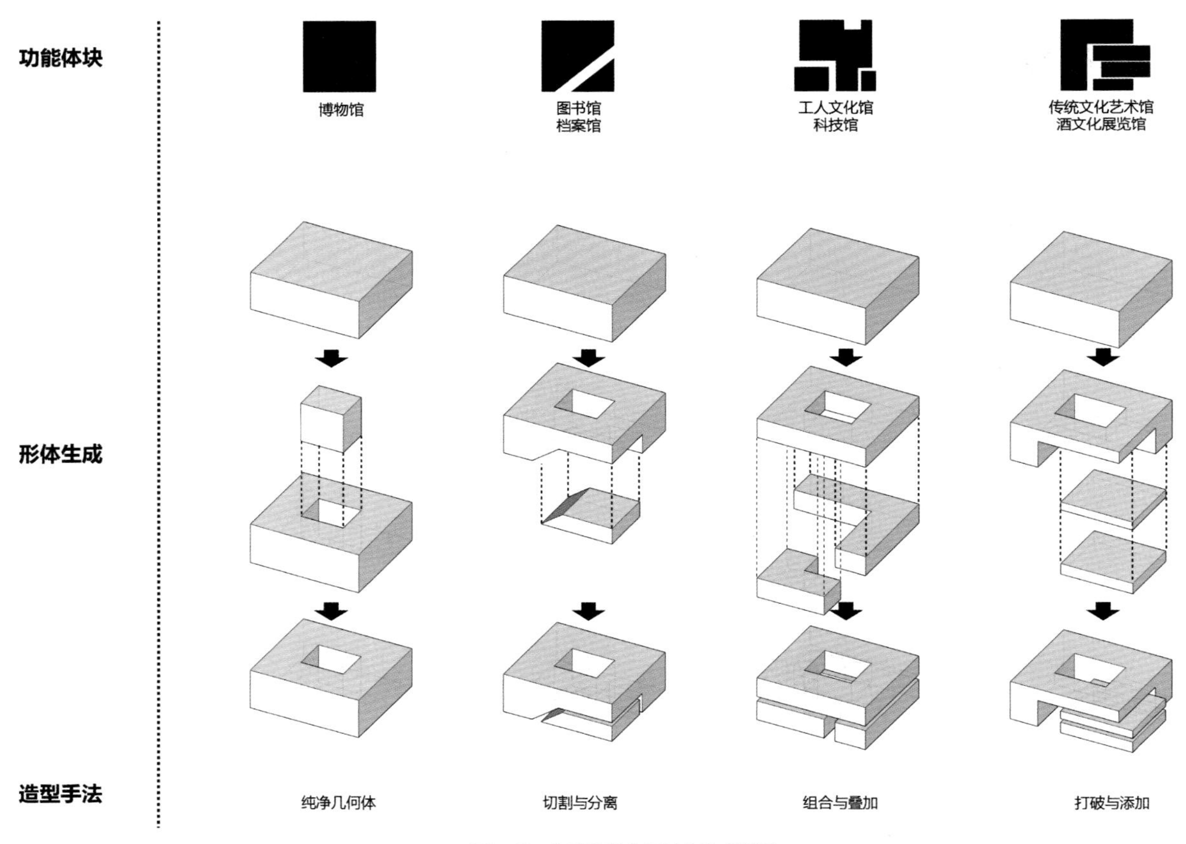

图3-16　各建筑单体设计的生成逻辑

3.2 空间同质化

基于当代建筑学多学科交叉的背景，空间同质化的特点是从其他学科，如生物学、物理学等的角度，来思考特定的建筑命题。

一般意义上来说，“均质”指的是均匀的质量或均匀的状态（相同或相似的性质），它也可寓意一个系统具有均匀的结构。在物理学中，“均质”用以描述一种材料或系统中，每一个点都具有相同的空间特性；它也用以描述物质或对象其属性不随位置的变化而变化。在建筑学科领域，这个原则也可相应地演变成一种设计逻辑，运用于特定语境的建筑创作当中。基于这种理念，建筑空间可以设计成由相同或相似的模块组成，并通过一定内在逻辑与秩序组合起来的生物系统，或是通过相同或相似的组织逻辑把不同模块组合起来的整体（图 3-17）。

空间同质化的设计策略通常需从“标准化”入手。也就是说，在设计过程中，首先需确定合适的尺度及方法，以把建筑空间规范成一系列标准化的单元，然后通过预设的结构及逻辑关系把这些空间单元有效地组织起来。

在这种空间模式中，合适的尺度及空间层次是关键；而其中系统化思维起至关重要的作用。相对于常规的从基地环境或功能形式出发的设计模式，系统化思维模拟自然界的生物特性，使建筑设计能与其他学科之间有更加全方位的协作与交叉，从而为建筑学开辟更广阔的设计思路。

3.2 Spatial Homogeneity

Based on the interdisciplinary thinking, spatial homogeneity tends to solve various design problems from the perspective of other disciplines, such as biology, physics, and so on.

In a general sense, homogeneity is defined as the quality or state of being homogeneity (of the same or similar nature). It also means having a uniform structure throughout. In physics, homogeneity usually means describing a material or system that has the same properties at every point of space; it also describes a substance or an object whose properties do not vary with position. Based on the concept of spatial homogeneity, architecture appears as a bio-system with uniform elements articulated by an inherent logic or as a system with different elements, but with a uniform composition.

This design approach is normally preceded by 'standardization'. During the design process, designers need to find appropriate scale and ways to standardize the architectural spaces into a system; and then, different spaces are arranged according to particular logic.

In this spatial model, systematic thinking is considered to be essential in terms of various hierarchies and scales. Other than conventional design methods beginning from programs and forms, systematic design opens up a much wider range of method-

这种空间模式下的自然化建筑，并不是在常规建筑设计上附加一些内容，而是成为设计过程中一个强大的核心，引导我们找出建筑内部的逻辑关系或空间现象背后的关联性，从而使建筑具有更加复杂的内涵及强大的适应力，就像自然界的生命体一样。换言之，自然化设计并不是仅局限于某些特定技术或建造过程，而是成为一种意识形态，包含对自然环境的态度和把设计付诸实现的一套方法。当常规的还原性和线性思维已不能够解决当代建筑学所面临的庞大复杂的能量流和参数变量等问题，系统化思维可以通过严谨和科学的规则形成自下而上的结构，从而具备解决复杂和非线性问题的巨大潜力。

ologies embedded with thoughtful collaboration between architects and other disciplines. In this sense, N-architectural design is no longer merely an addition to certain boxes that architects have already made, but it truly becomes a powerful and meaningful agenda to figure out internal logics or connections behind spatial phenomena. Systematic thinking is not just a word defined by certain technology or construction; it is an ideology including attitudes towards our world and the methodology towards design. In other words, conventional processes with reductive and linear thinking are not able to solve the problems in contemporary architecture facing so many complex flows and variables, whereas systematic thinking with rigid and robust rules forms a bottom-up structure and has higher potential to deal with complicated and non-linear situations.

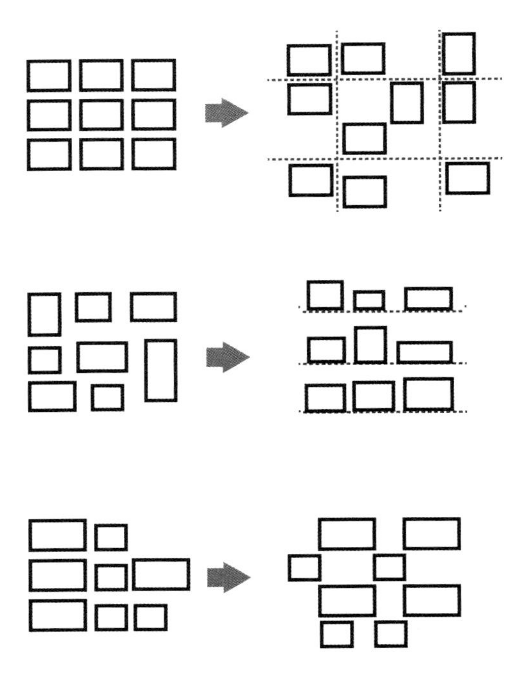

元素相同，逻辑不同

元素不同，逻辑相同

元素混合，逻辑混合

图3-17　同质化空间类型

3.2.1 艺・墟

“艺・墟”原本是一个关于艺术博物馆的设计。设计场地位于城市的一个边缘地带，发展之前主要是郊野农田，现正被一些新兴工业建筑所包围。由于场地所处区域尚未高度城市化，聚居的主要是一些来自其他城市和处于创业发展阶段的年轻群体。因此，这是一个人工与自然、城市与景观相互竞争的区域。通过对环境的思考和分析，设计师最终得出结论，在这里建造一个艺术博物馆其实并不合适；相反，构建了一个艺术社区，为尚处于探索阶段的年青艺术家提供一个工作、展示，并能与公众互动交流的场所，是符合社区需求并能激发社区活力的一种方向。

基于这种理念，设计首先在基地内导入 10 m×10 m 的网格，形成均质的空间组织逻辑。而功能空间根据不同艺术形式，发展出以 10 m×20 m 为基础的标准化结构模块，形成一系列能满足不同使用要求的空间。然后，通过扭转、重叠等设计手段，产生一系列各具特色的艺术空间和立体化的庭院系统。在这个设计中，简单的网格逻辑不仅实现各要素之间的结构性衔接，也创造出适合不同艺术形式使用的空间类型。此外，通过空间的同质化模式，建筑很好地实现了与自然的融合。系统化设计思维的力量在此设计中得到充分体现（图 3-18~ 图 3-22）。

3.2.1 M+Arts

According to the original design brief, a conventional art museum was planned to be built on the site. The site is situated at the edge of the urban area, formerly covered with farmlands but currently surrounded by a number of industrial buildings. Not having experienced intensive urbanization, the area where the site is located is mainly inhabited by a large number of youngsters, most of whom migrated from regional areas and are still striving for their future development in the city, and thus this is a place full of conflicts and tensions. In this sense, to provide sufficient development opportunities for these young people is defined as a problem that the design aims to solve. Based on the analyses, the designer pointed out that an art community rather than an art museum should be built in the area. The art community not only will provide a place for emerging young artists to work, to exhibit, and to exchange ideas, but also a platform to encourage the interaction between young artists and the public.

In light of the design concept, a 10 m×10 m grid system is then introduced into the site, forming a new spatial organizational logic. Also, the studio spaces are classified and further developed into 10 m×20 m modules. And then, through twisting, overlapping and other design strategies, a typology of spaces and a courtyard system are finally produced to meet the use requirement of different forms of art. From this design work, it is noticeable that the spatial logic set out by the grids appears simple; however, it not only produces flexible spaces for different artists, but also achieves coherence and continuity among various elements. In this sense, spatial homogeneity, deemed a mechanism, combines all spatial elements into an organic entity, to be well integrated with nature. The strength of systematic thinking has been fully manifested in this example.

图3-18a　城市与自然景观的对峙

Located in rural-urban continuum, the site is facing erosion of urbanization. We want to explore the possibility of maximum communication between ARTIFICIALIZATION and NATURE. Building form miniatures the clash of urban and nature,while in terms of space,building and landscape interpenetrate with one another.

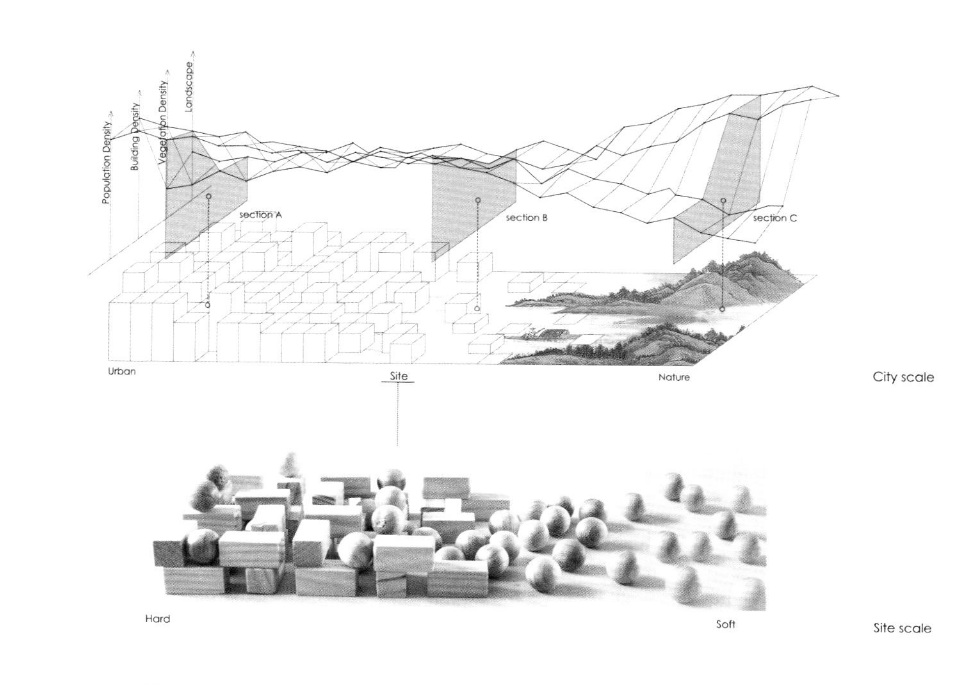

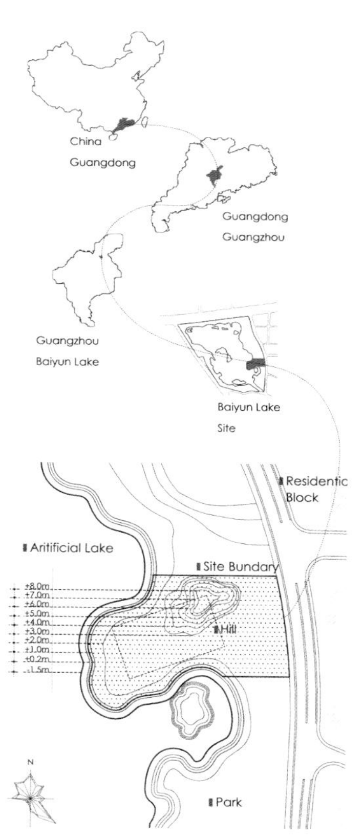

图3-18b　地形的分析与营造

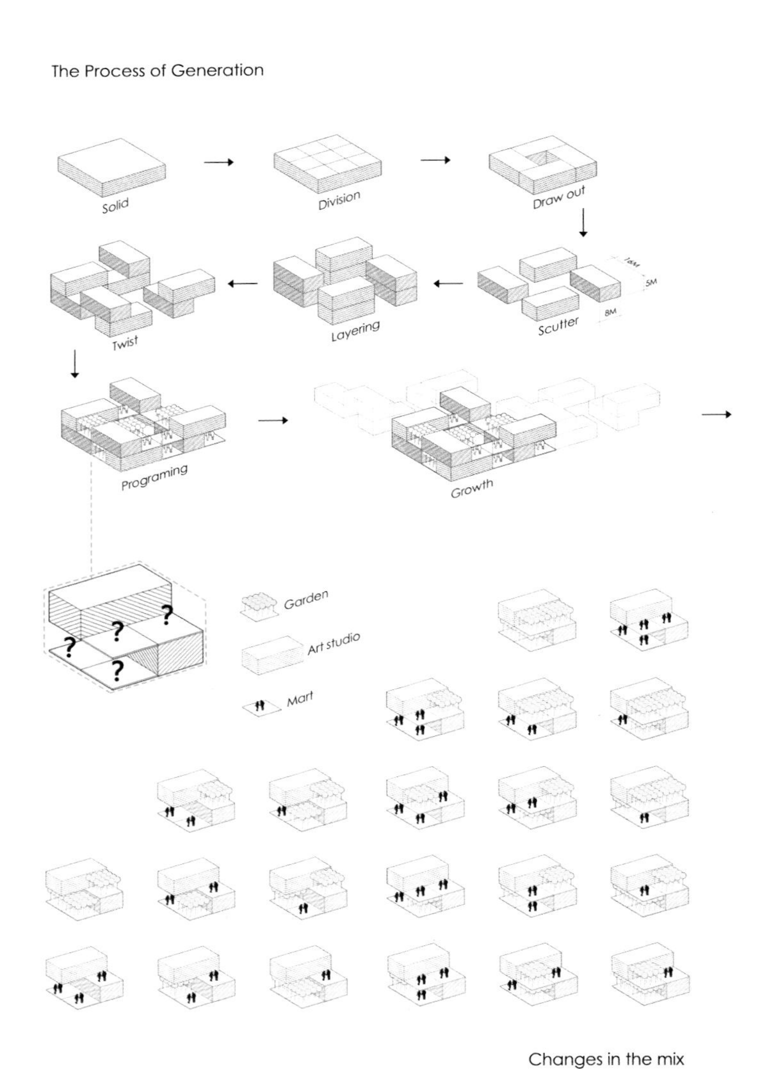

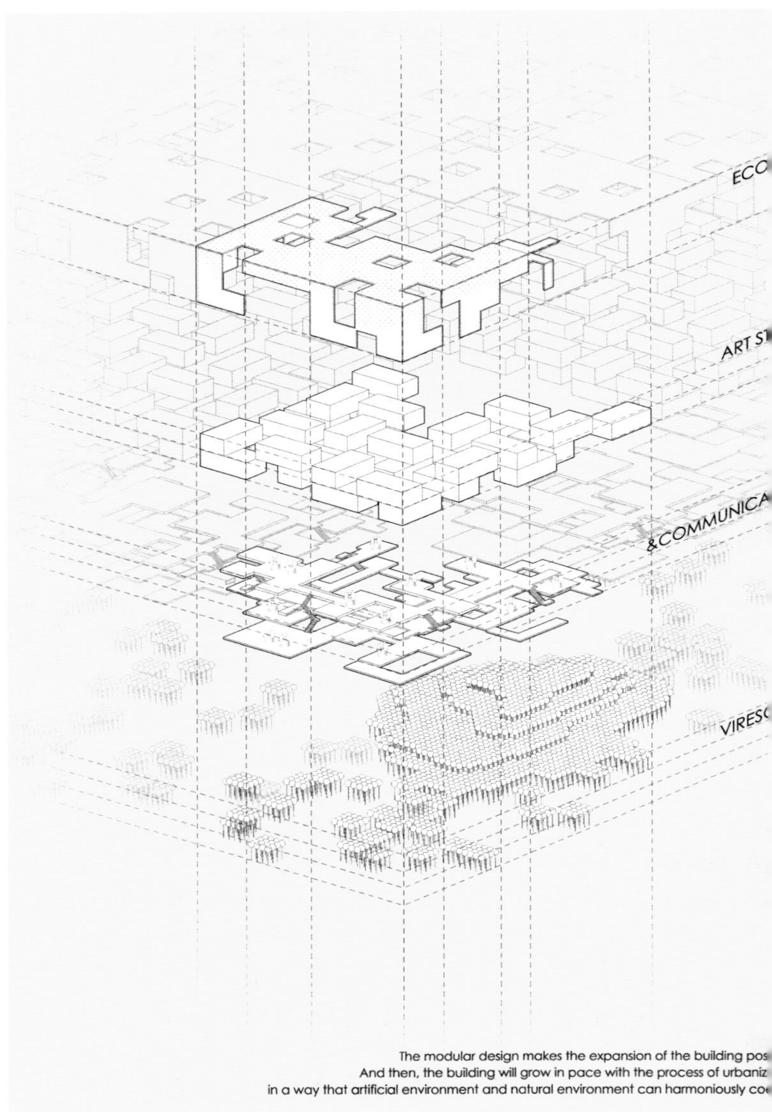

图3-19　空间模块的生成与组合逻辑

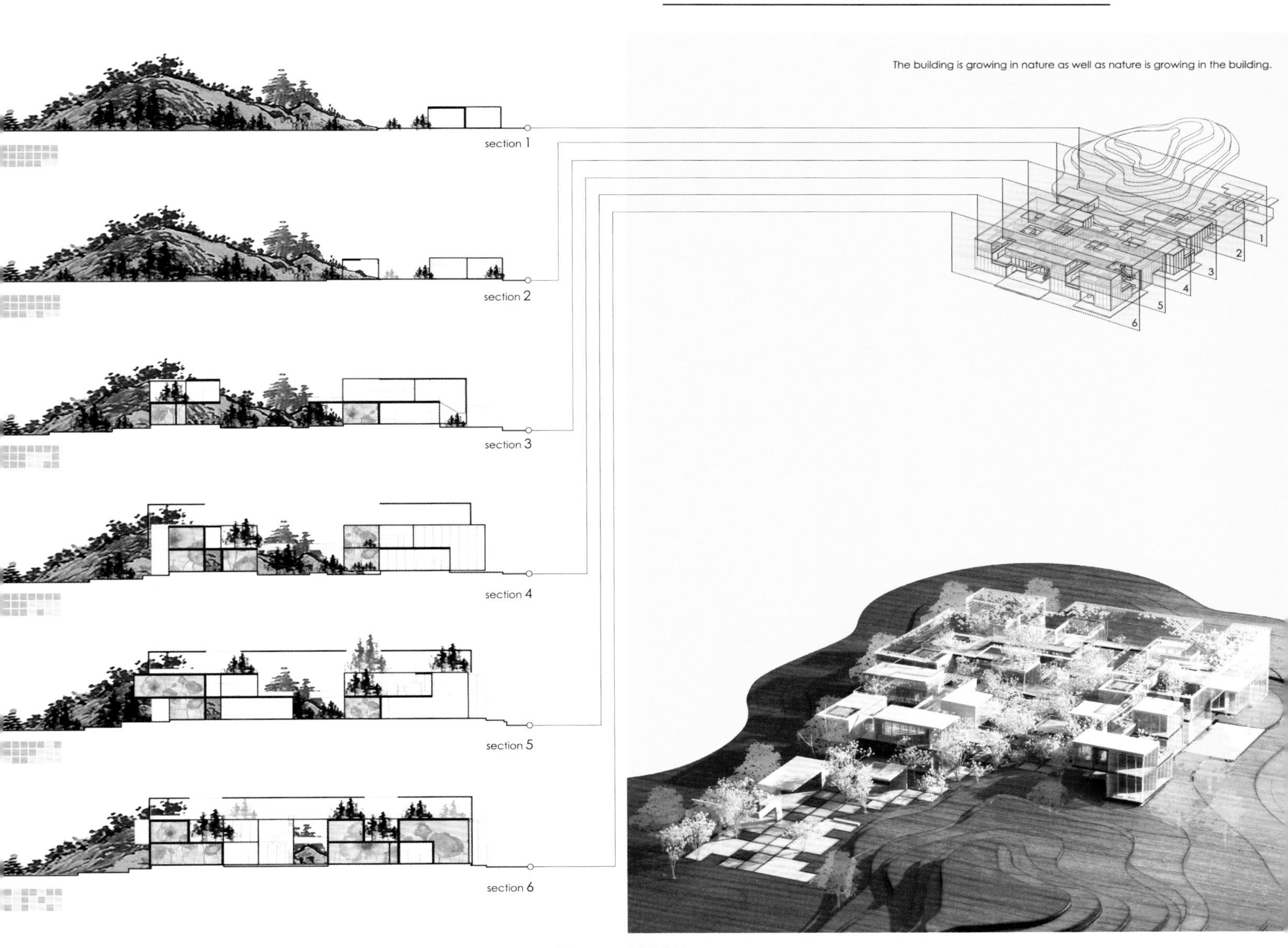
The building is growing in nature as well as nature is growing in the building.
section 1
section 2
section 3
section 4
section 5
section 6
1
2
3
4
5
6

图3-20　剖面分析

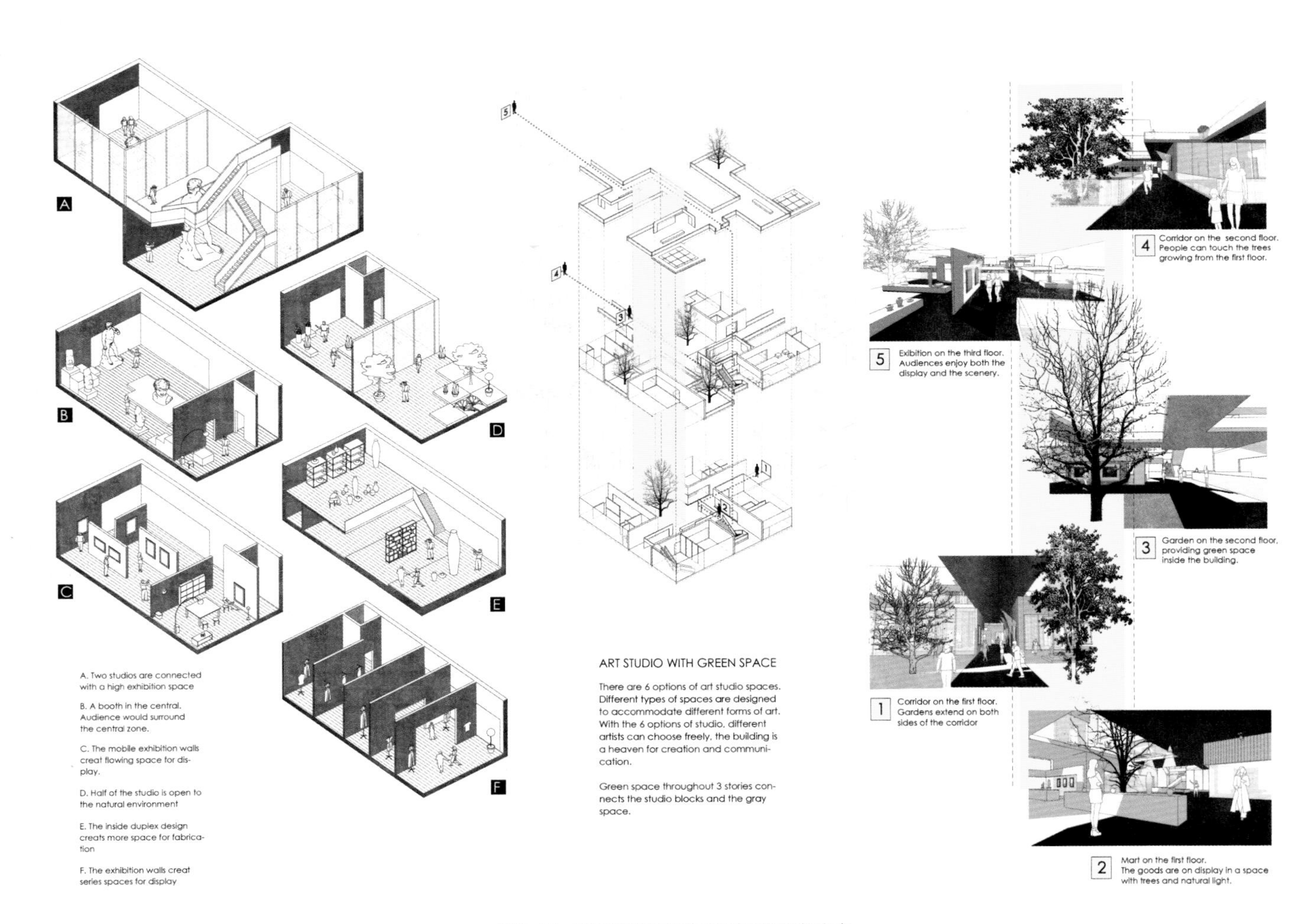

图3-21　不同类型艺术空间及庭院系统的产生

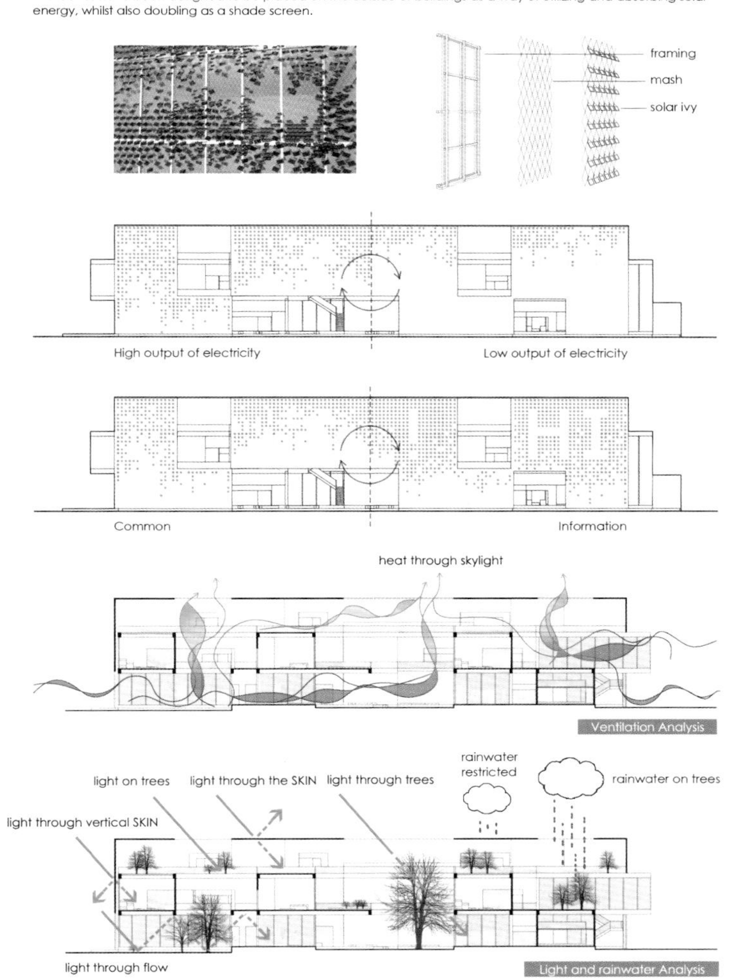

图3-22　气候适应性生态设计

3.2.2 “裂 • 变”：湿地生态博物馆

“裂 • 变”的设计驱动力也是来自对周围环境的研究与思考。场地处于城市的边缘地带，原本是颇具岭南特色的郊野农田。设计因此引入裂纹肌理，延续原有农田景观记忆，同时以裂纹肌理组织建筑空间，从而使建筑成为自然景观体系的有机组成部分。裂纹系统，既是建筑形式，也是组织交通流线和景观要素的结构逻辑，建筑空间因而与景观浑然一体。由此可见，通过同质化的空间手段，建筑也可轻易实现与自然的一体化，使人们能在探究自然的过程中获得自然的熏陶及精神上的升华（图 3-23~ 图 3-31）。

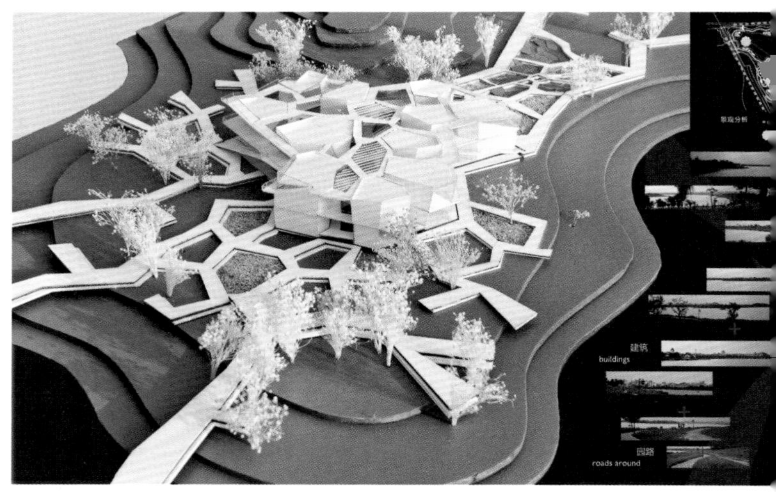

图3-23 湿地博物馆设

3.2.2 Crackle: A Wetland Ecological Museum

The design concept was initiated from the analysis of the surrounding environment. The site is located on the edge of the city and was originally covered with farmlands. In order to maintain the continuity of the surrounding farmlands and to make the building an integral part of the natural landscape system, a crackle pattern was introduced into the site. Based on the crackle pattern, various functional spaces were organized and linked by a three-dimensional circulation system. In this example, the crackle pattern is not only a symbolization of nature, but also functions as an eco-system so that architecture can be completely dissolved into the landscape. As a result, people are able to experience the nature and be inspired by the nature while visiting the museum.

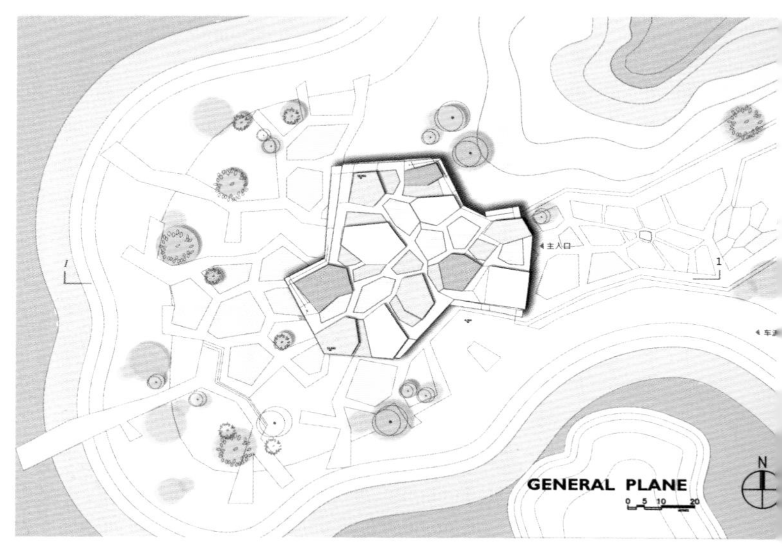

图3-24 湿地博物馆总

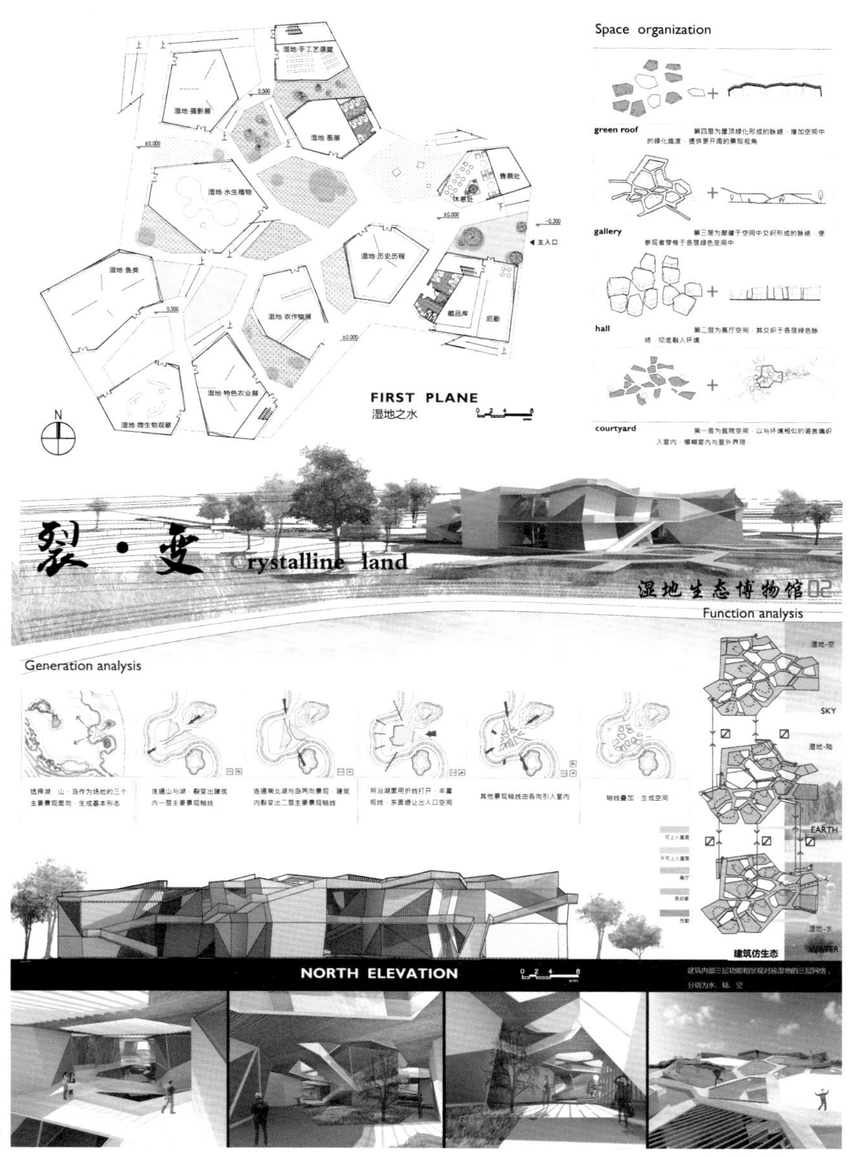

图3-25　湿地博物馆首层平面及流线分析图

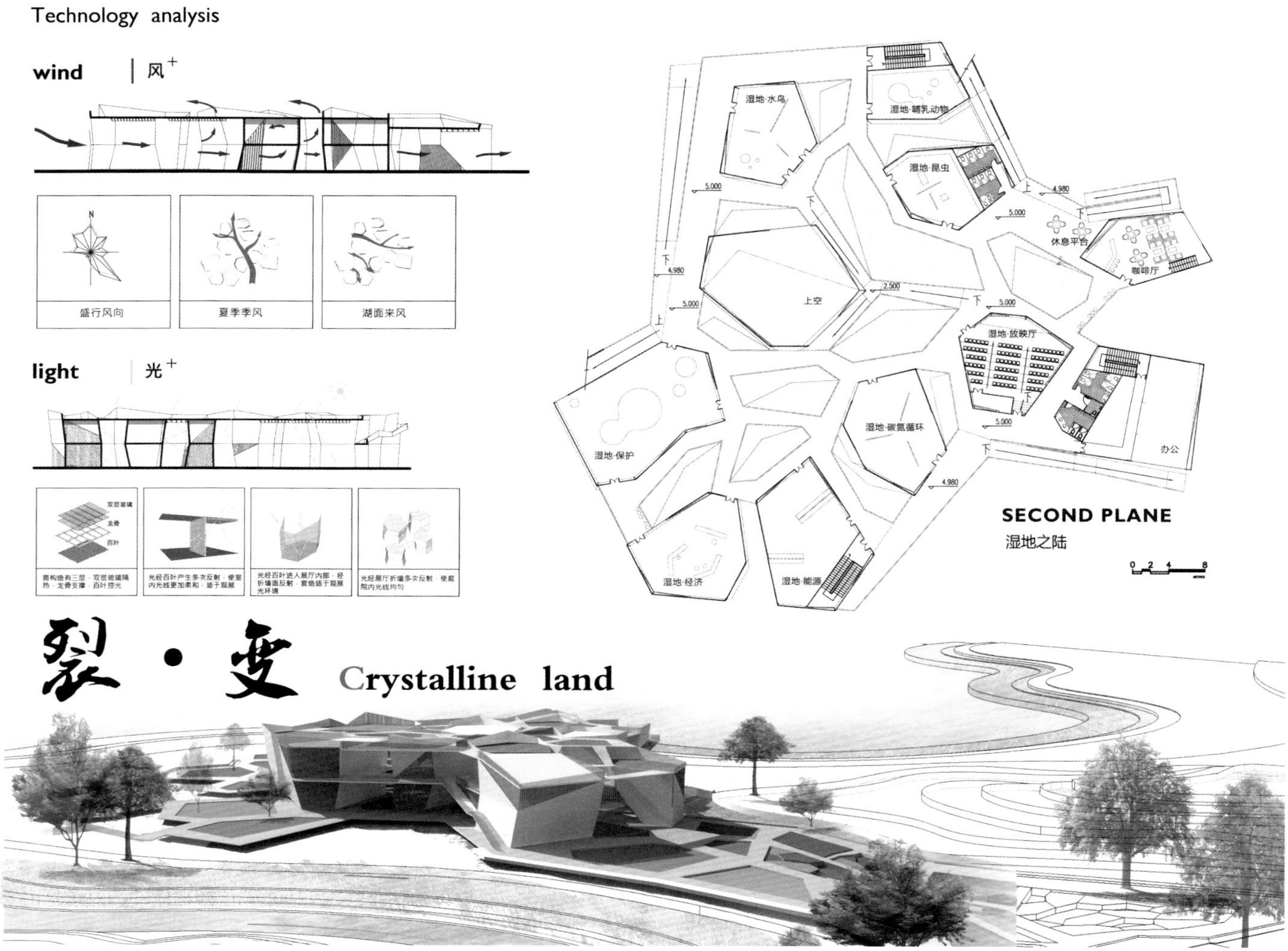

Technology analysis

wind | 风$^{+}$

盛行风向

夏季季风

湖面来风

light | 光$^{+}$

双层玻璃

龙骨

百叶

窗构造有三层 · 双层玻璃隔热 · 龙骨支撑 · 百叶控光

光经百叶产生多次反射 · 使室内光线更加柔和 · 适于观展

光经百叶进入展厅内部 · 经折墙面反射 · 营造适于观展光环境

光经展厅折墙多次反射 · 使庭院内光线均匀

湿地·水鸟

湿地·哺乳动物

湿地·昆虫

休息平台

咖啡厅

上空

湿地·放映厅

湿地·碳氮循环

湿地·保护

办公

湿地·经济

湿地·能源

SECOND PLANE

湿地之陆

裂·变 Crystalline land

图3-26 湿地博物馆技术设计图

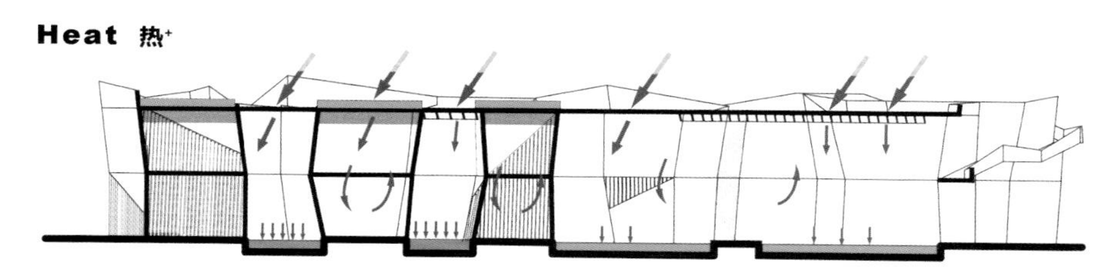

夏季百叶控制热量的进入，而建筑内部的植被和水有效调节室内温度

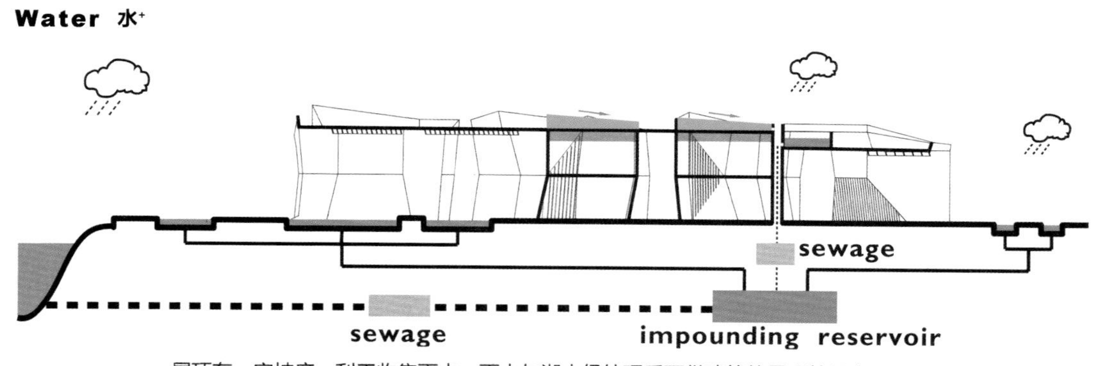

屋顶有一定坡度，利于收集雨水。雨水与湖水经处理后可供建筑的景观等用水

图3-27　湿地博物馆技术设计图

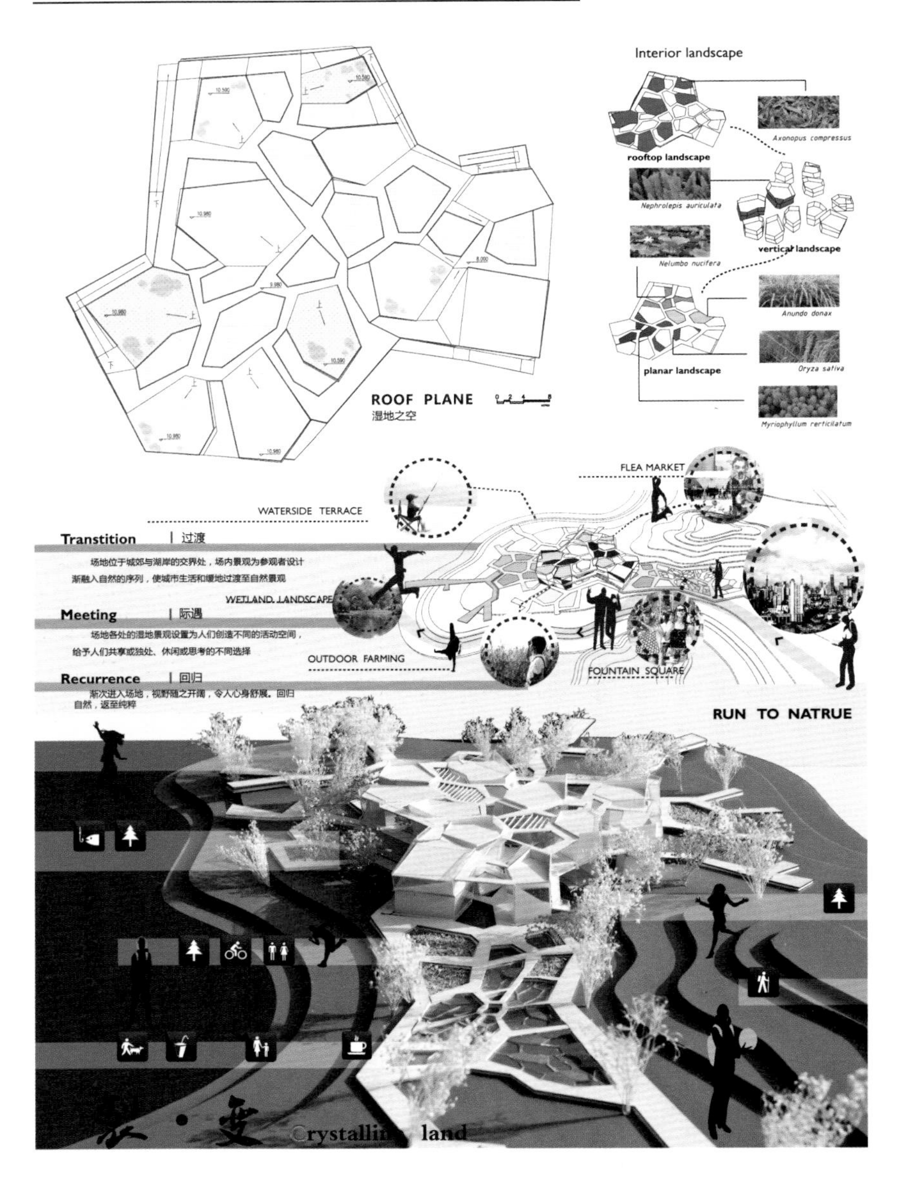

图3-28　湿地博物馆景观分析

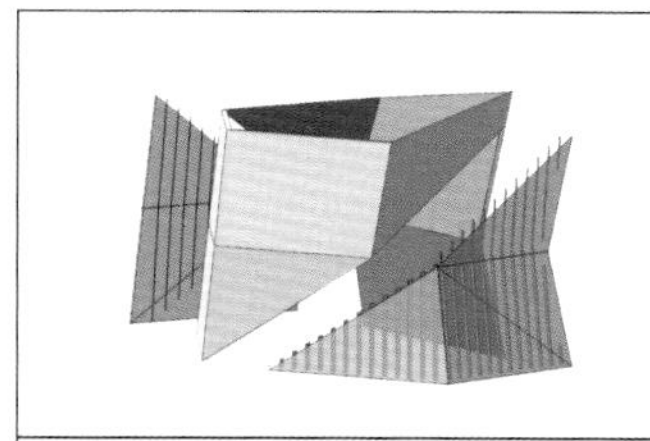

外围护采用玻璃、百叶玻璃及实墙三种形式，增加建筑虚实空间层次的变化

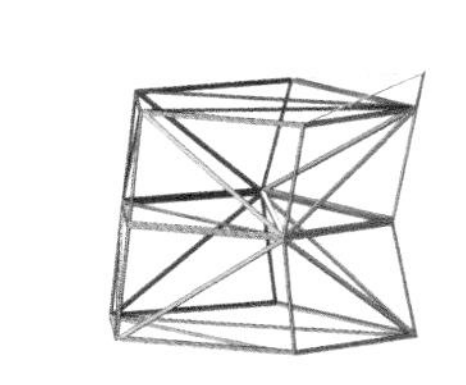

采用钢结构作为建筑骨架，构造形式与立面相适应

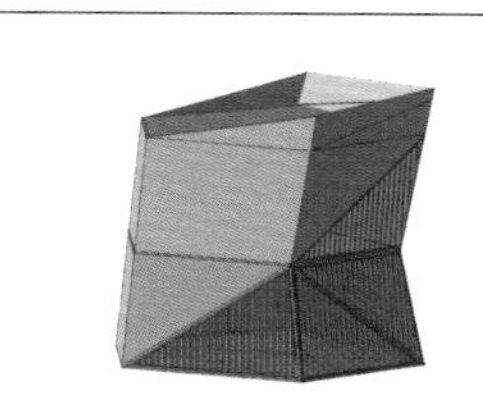

展厅建筑单元形成

图3-29　湿地博物馆结构分析及模型照片

图3-30　湿地博物馆模型照片

图3-31　湿地博物馆模型照片

3.2.3 “多孔体”：岭南艺术博物馆

“多孔体”是自然界中存在的一种同质化现象，而该艺术博物馆设计是受多孔体启发，并以其结构逻辑来解决相关建筑命题的。首先，设计根据使用要求对功能空间进行梳理，创建一系列标准化的空间模块，使每个模块能容纳不同规模的展览用途；然后，设计根据一个立体化的网络系统对模块进行叠加和组合，形成建筑整体。同时，通过抽减的方法，在模块之间形成丰富的“孔状”空间，用以布置公共交流空间及组织交通。孔状空间不仅引入阳光、空气，同时也把绿化景观带入建筑内部，从而使自然景观要素成为建筑整体结构的一部分，有效地实现建筑自然化表达（图 3-32~ 图 3-36）。

3.2.3 Porosity: Lingnan Art Museum

Porosity is a phenomenon of homogenization commonly found in nature. Being inspired by the principle of porosity, this art museum design used its structural logic to solve particular architectural problems.

At first, based on the requirement of the programme, the entire museum was developed into a series of spatial modules in uniform scales and structure, with each module accommodating different sizes of exhibition space. In light of the principle of spatial homogeneity, a three-dimensional grid system was introduced into the site, and then the spatial modules were articulated and stacked to form a cube, with various pores between the modules functioning as public spaces or circulations. The pores not only allow natural light and ventilation into the museum, but also bring inside the surrounding greenery to form multi-level gardens, effectively realizing the naturalization of architecture.

图3-32 岭南艺术博物馆模型

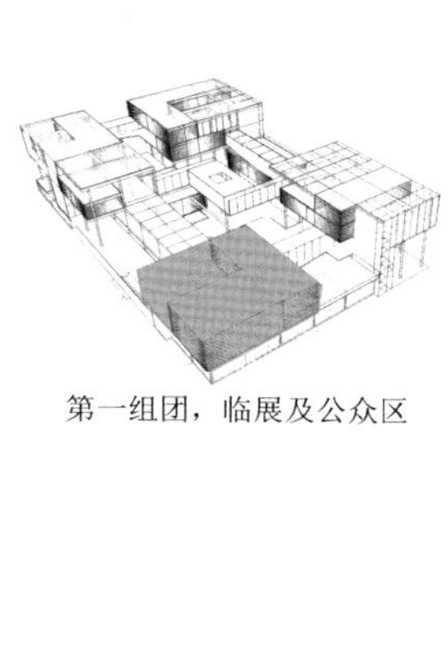

第一组团，临展及公众区

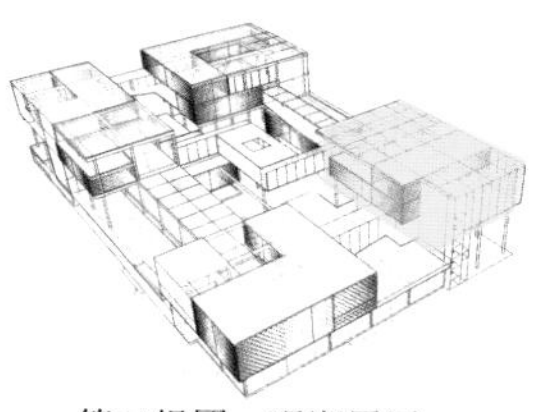

第二组团，观湖展区

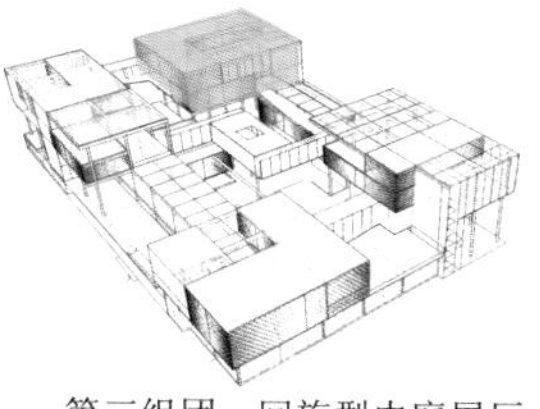

第三组团，回旋型内庭展厅

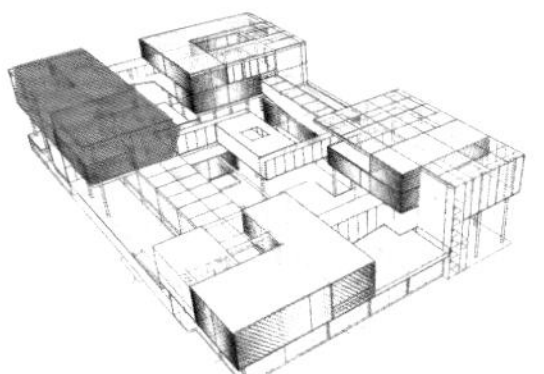
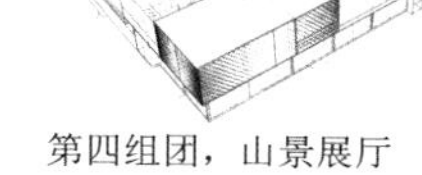

第四组团，山景展厅

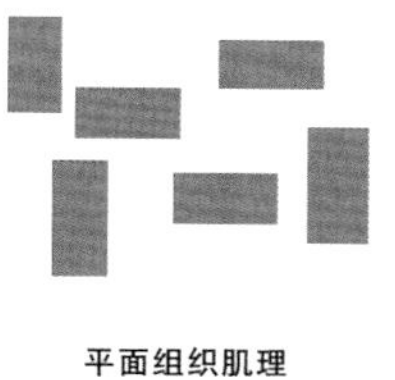

平面组织肌理

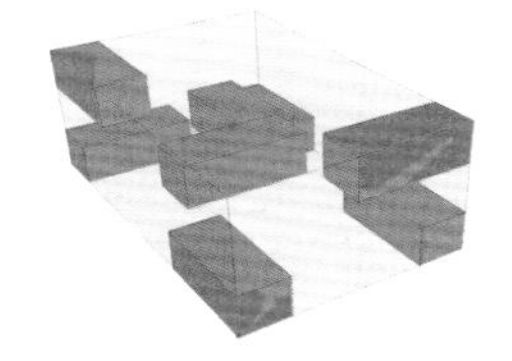

立体空间组织逻辑

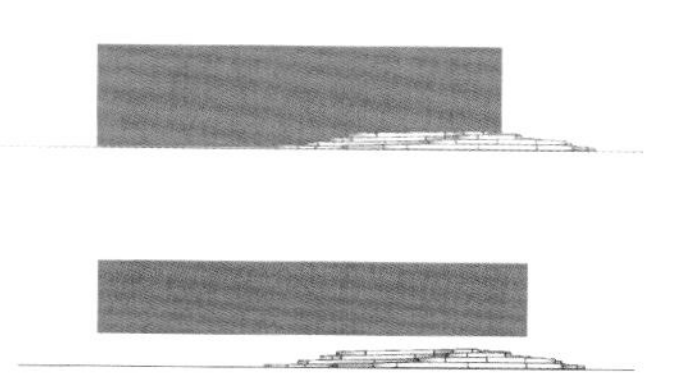
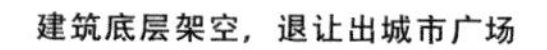

建筑底层架空，退让出城市广场

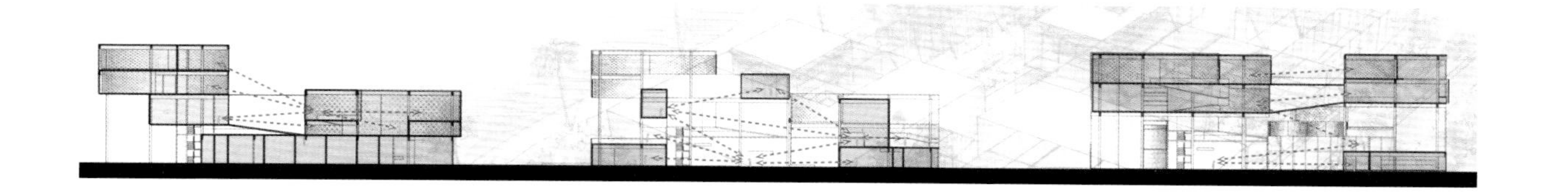

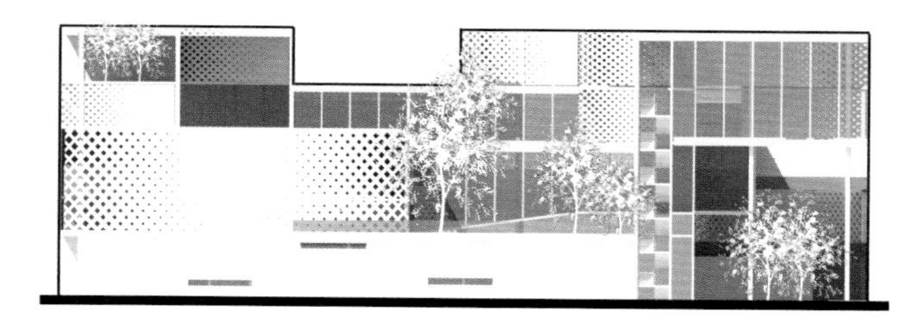

图3-33　岭南艺术博物馆设计概念剖面图分析(上)及立面图(右)

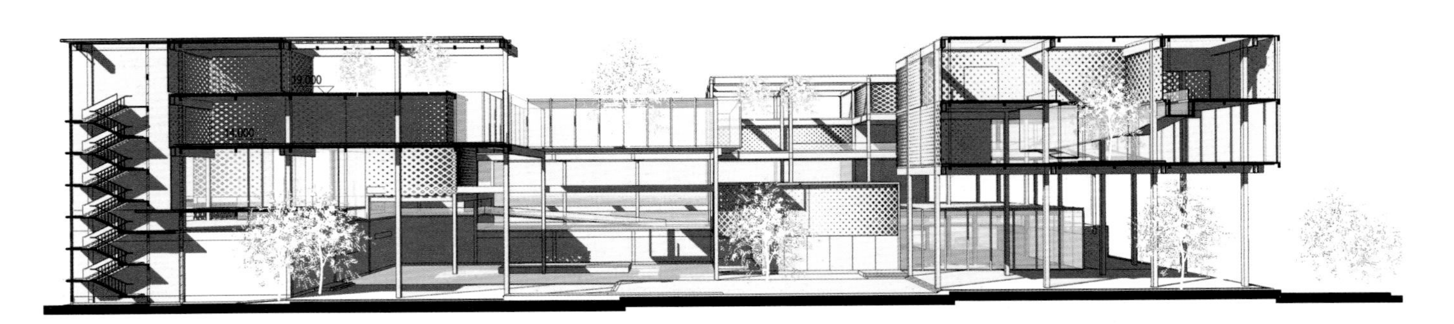

图3-34　岭南艺术博物馆剖面图

图3-35　岭南艺术博物馆模型

图3-36 岭南艺术博物馆模型

3.3 生态社会

生态社会是基于自然化建筑是解决特定社会问题的方法之一这一理念而产生的设计类型。该设计模式主要探讨自然如何成为一种契机或触媒，以促进社会公平、平等与和谐，并实现社会的可持续发展。由于人类拥有趋向自然的本质，设计通过提供人与自然互动的机会，净化人的心灵，达到精神重塑的目的。同时，通过构建建筑与自然的关系，也可以提供一个自然化的空间场所，重塑人与人之间的关系与社会秩序，帮助创建一个理想的社会结构。这也许可以帮助解释为什么自然景观在当代设计实践中已成为重要的主题和符号。

3.3 Eco-society

Eco-society is a design approach based on the concept that N-architecture is one of the methods or means to solve particular social problems. This design model mainly discusses how nature can be an opportunity or a catalyst to promote social fairness, equality and harmony, helping achieve social sustainability. Owing to the inherent characteristics of human beings, through providing opportunities to interact with nature, human minds are able to be purified and their spirit and value are also transformed. Though establishing the relationship between nature and architecture, N-architecture can help rebuild people's relationship so as to create an ideal social structure. This somehow can help explain why natural landscape has become an important notion in contemporary architectural design.

3.3.1 “垂直社区”：都市附加住宅

都市附加住宅旨在为香港等高密度城市寻求可持续性的替代发展模式。这是一个对都市废弃的采石场进行再利用、重塑自然化社区的设计。设计从古代悬空寺和传统穴居模式中吸取灵感，通过把周边城市肌理从水平方向延伸至垂直界面，创建一个别具一格的“空中社区”。另外，通过在不同层面创建公共交往平台和提供自动化步行交通系统，设计有效地实现人与人的互动和交流，并塑造和谐的社区氛围。与此同时，绿化景观也沿平台延伸至社区内，为人们提供接近自然的机会。藉着这个方案，我们得到的启示是，当高密度已成为香港市民不可避免的生活方式，垂直社区开辟了多维度发展城市空间的新思路（图3-37~图3-42）。

3.3.1 Vertical Community: Urban Housing Plus

This project aims to seek an alternative development model for high-density cities like Hong Kong. In this case, an abandoned quarry site within the urban area was reused and revitalized. Being inspired by Chinese ancient cantilever temples and cave dwellings, the design created a vertical community by extending the surrounding urban fabric from the horizontal to the vertical interface. Through providing platforms and escalators at different levels, the integration among occupants was well achieved.

Meanwhile, the surrounding greenery has also been brought into the community along the platforms in order to enhance people's access to nature on a daily basis. From this conceptual design, it is noticeable that while vertical living has become an inevitable lifestyle of Hong Kong citizens, the vertical neighbourhood has opened up a multi-dimensional development method for high-density built environment.

场地位于都市废弃的采石场

设计通过创造垂直社区，旨在对废弃空间进行再利用。

图3-37 场地原貌与“垂直社区”的植入

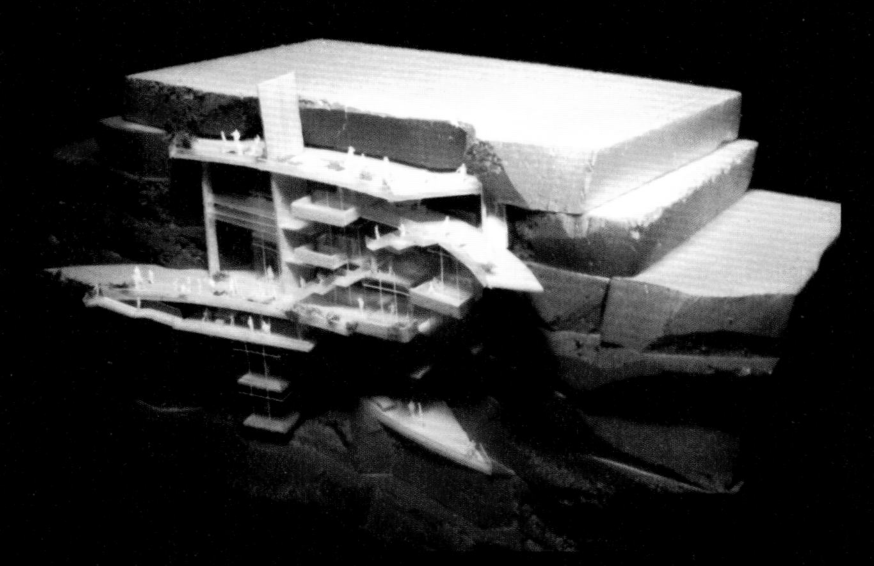

垂直社区的单元模型

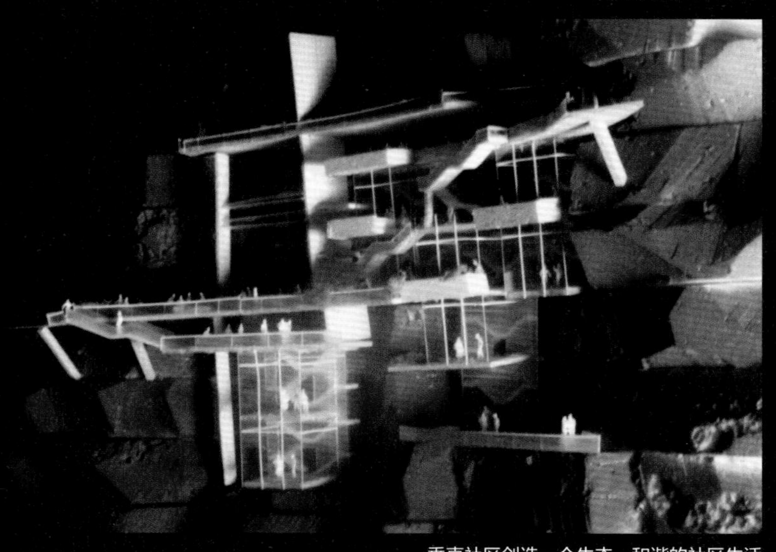

图3-38 “垂直社区”单元模型

垂直社区创造一个生态、和谐的社区生活

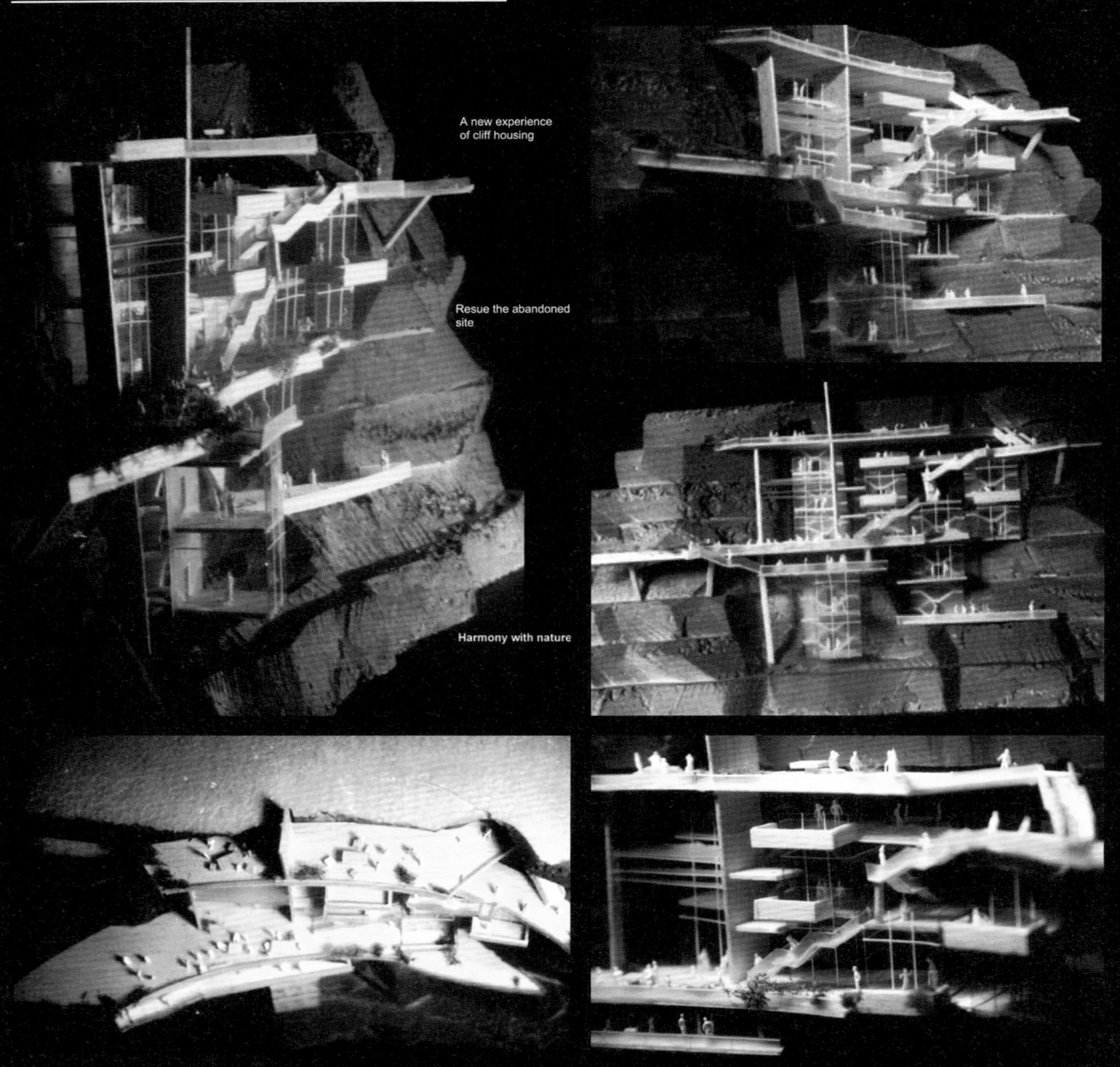

图3-39 “垂直社区”设计概念模型

图3-40 “垂直社区”设计概念模型

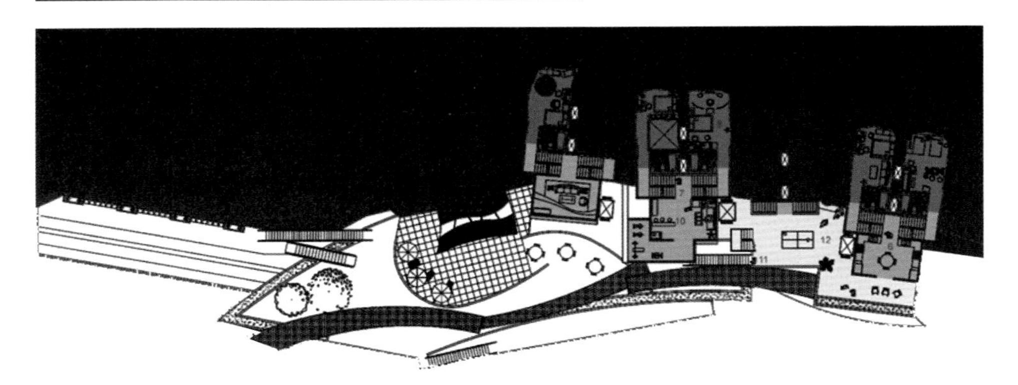

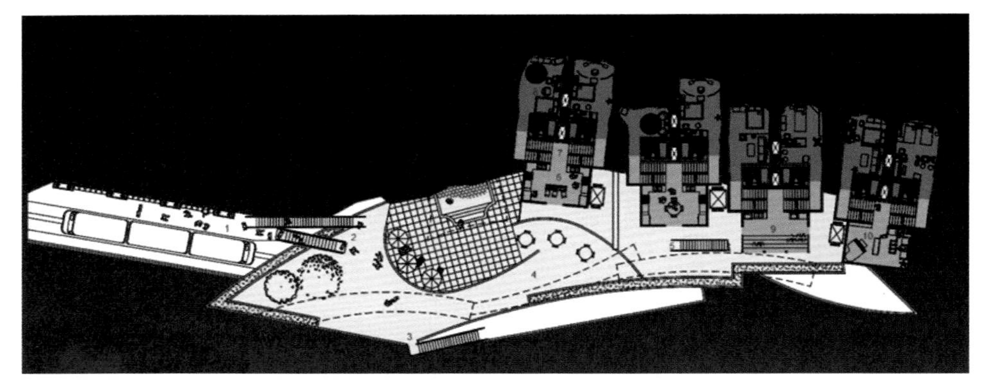

图3-41　社区单元平面图

图3-42　社区单元立面图

3.3.2 “流动花园”：高校图书馆设计

藉着对周边环境的分析以及对学习环境与自然环境之间关系的思考，设计提出发挥自然景观的教育功能的设想。因此，基地周围丰富的绿化景观成了建筑概念生成的驱动力，通过在基地内引入两条交叉轴线，形成建筑空间的组织架构，在分割不同功能区域的同时，也把绿化引入建筑内部并延伸至屋顶界面，形成一个贯穿整体的“流动花园”。花园结合学生的步行活动系统，在图书馆内部创造一个能持续与自然互动的学习环境，很好地实现了建筑环境生态化和生态环境教育化的设计意图（图 3-43~ 图 3-45）。

3.3.2 Flowing Garden: A University Library Design

Initiated from the analysis of the context and research on the relationship between the people and nature within a learning environment, the designer raised a question as to how landscape can play a role in the educational process. As a result, the surrounding landscape pattern was carefully analyzed and treated as a driving force to generate design ideas. Through introducing two intersecting axes into the site, the library aims to establish a coherent relationship with the surrounding buildings. Functioning as a frame, the two axes divide the building into different functional areas, and also bring the surrounding greenery from the outside to the inside, vertically up to the roof, forming a 'flowing garden'. Combined with the pedestrian movement system of the building, the flowing garden has created a vivid place for students to interact with nature constantly while learning, achieving the ideas of building up an ecological learning environment.

图3-43 “流动花园”设计模型

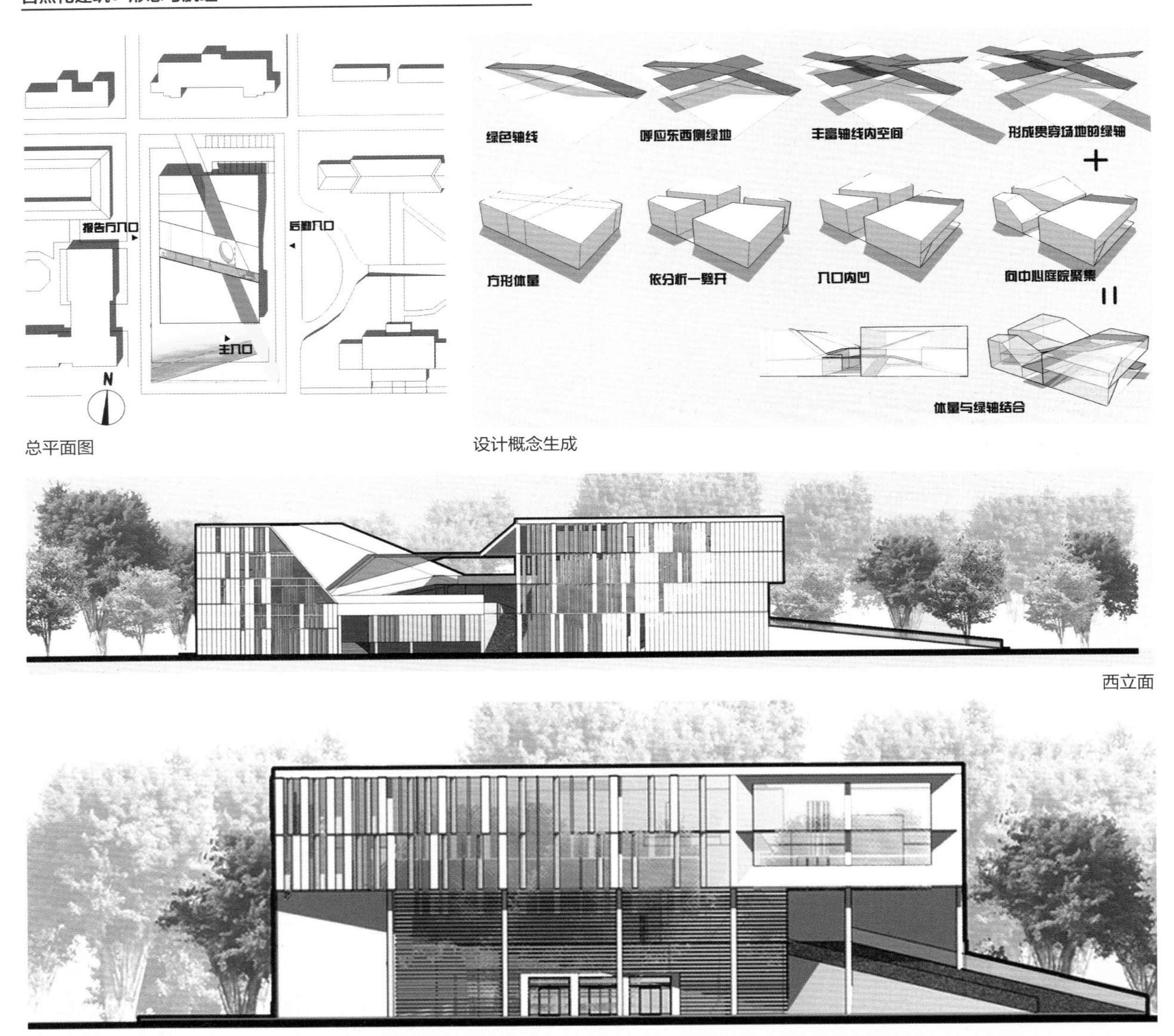

总平面图

设计概念生成

西立面

南立面

图3-44　“流动花园”立面图

图3-45　“流动花园”设计模型

3.3.3 “折叠与起伏”：岭南艺术博物馆

生态社会，就是人与自然的和谐，是传统社会一个基本的生活观念；因此，自然作为一种设计要素在乡土建筑中并不罕见。“折叠与起伏”是一个位于湖边地带的艺术博物馆设计。设计首先深入分析地域历史和文化，提炼出岭南文化多样和融会的特点；同时，结合对岭南建筑环境及其气候适应性设计特点的分析，总结出其具有多样和多层次的空间特点。基于这些思考，设计师构思一系列把自然引入建筑的方法。通过对一个连续界面的折叠与起伏，形成一系列高低起伏并与庭院穿插互动的艺术空间，从而为参观者营造出独特的空间体验（图 3-46~ 图 3-51）。

3.3.3 Folding and Undulating: Lingnan Art Museum

The idea of eco-society means establishing a harmonious relationship between people and nature, which was the basic living philosophy in traditional society. Therefore, nature as a design element is commonly found in vernacular architecture. 'Folding and undulating' is an art museum design along a lake. Other than analyzing the cultural significance of the region and the diverse and multivariate characteristics of Lingnan's vernacular architecture, the designers also put their focus on the environmentally and climatically responsive design features of local architecture and finally developed a series of strategies to bring natural elements into the museum. Through folding and undulating, a continuous interface has formed a series of functional spaces interspersed with gardens, in addition to creating a unique experience for visitors.

图3-46　“折叠与起伏”的设计模型

图3-47　“折叠与起伏”的设计模型

自然化建筑：形态与肌理

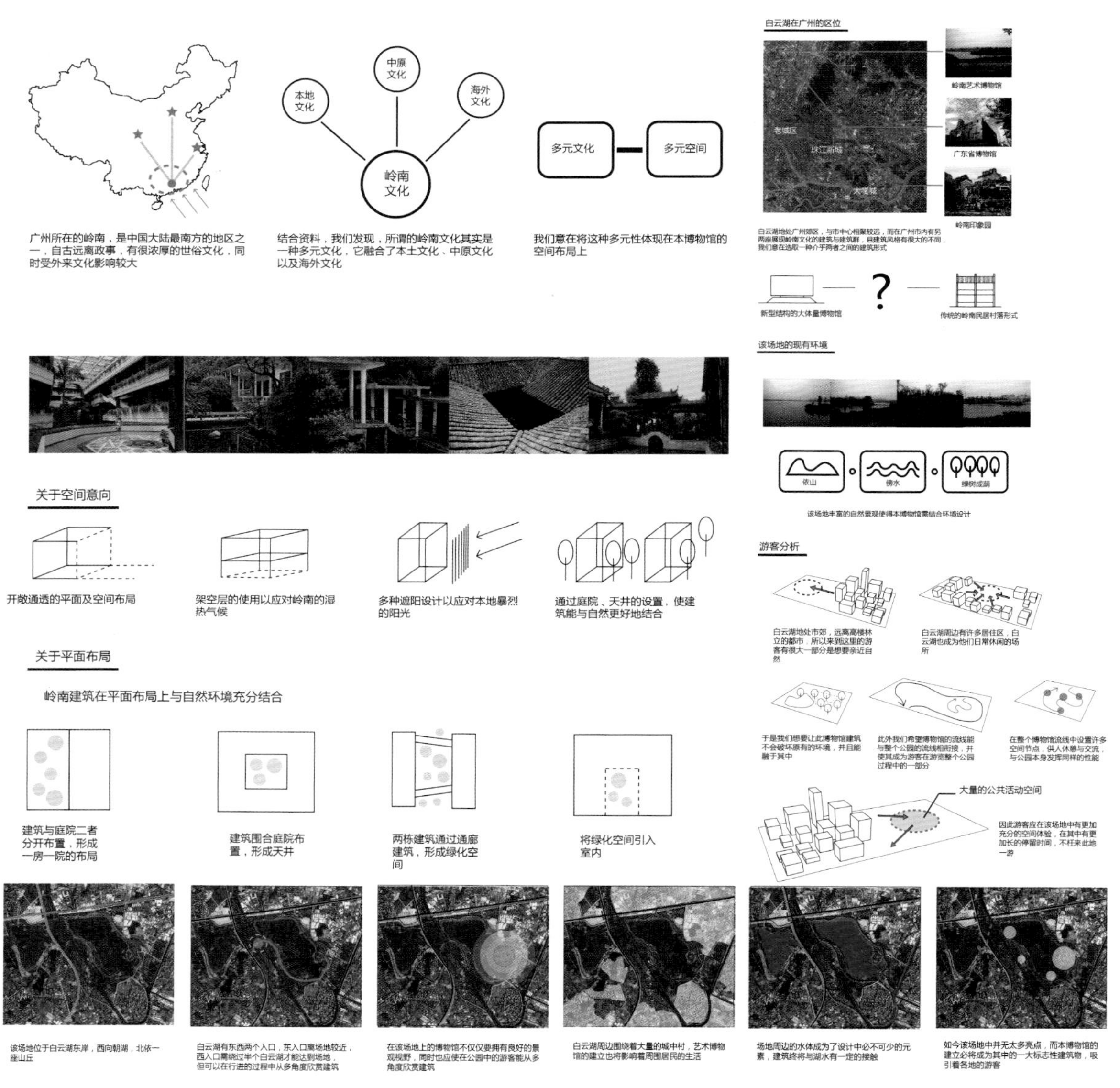

图3-48 “折叠与起伏”对于岭南文化的思考与场地分析

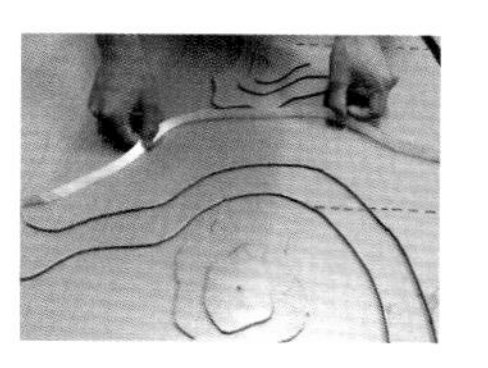
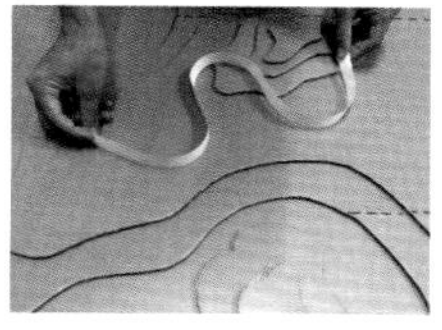
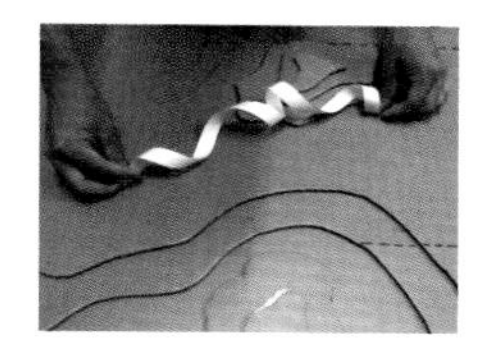
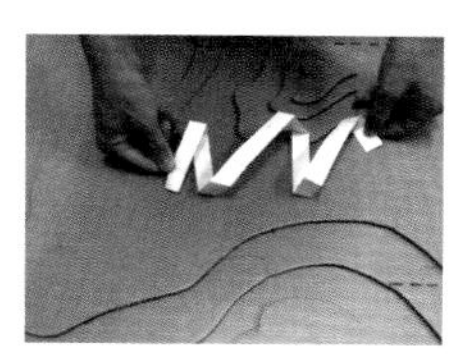

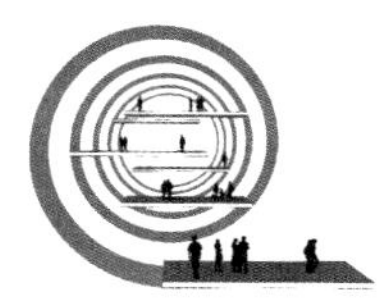
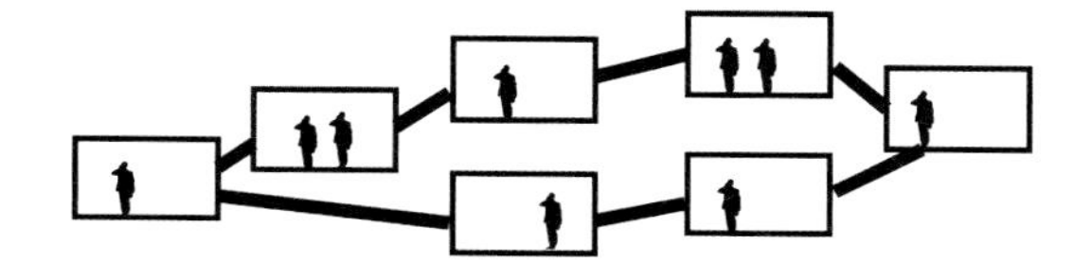

相比于上述这些形式均仅仅考虑内部空间的贯通，螺旋形的造型既满足了前者又能够使流线与山地和建筑造型相结合

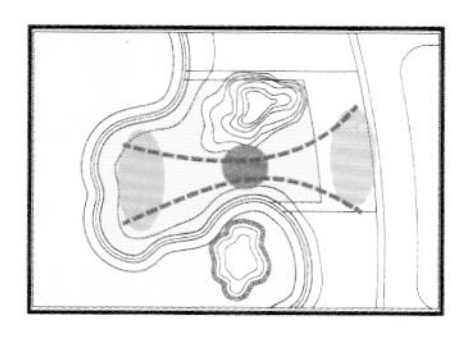

该场地由于有一座山丘，所以使东西流线受限

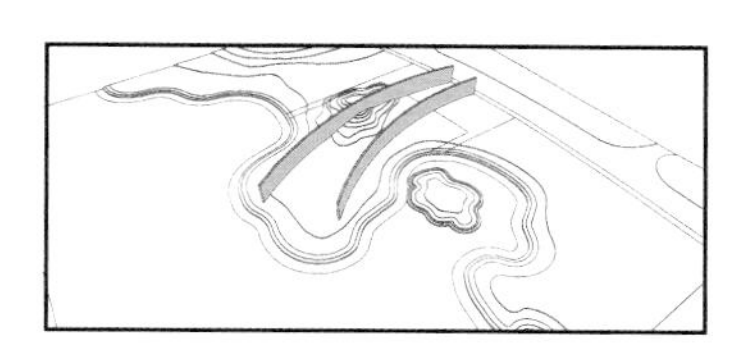

建筑由两条曲线控制顺着场地的弧度向西延伸

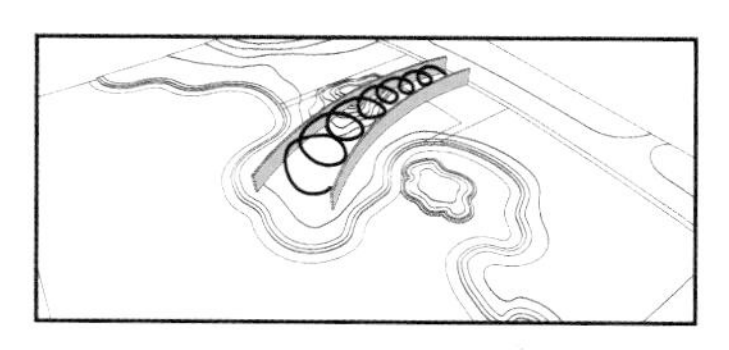

螺旋形的造型可以使建筑架起在山丘之上，转向湖边

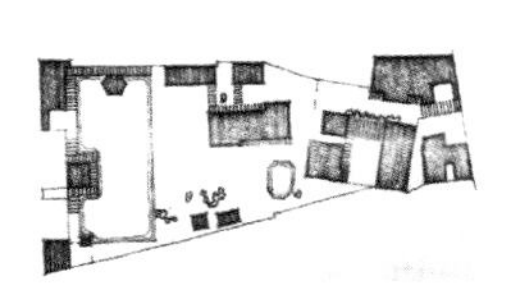
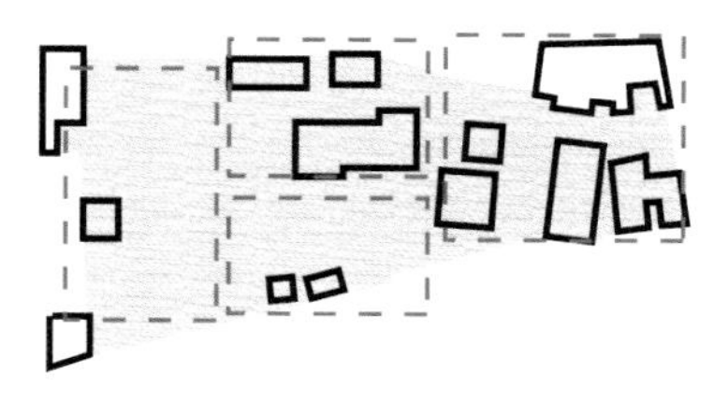

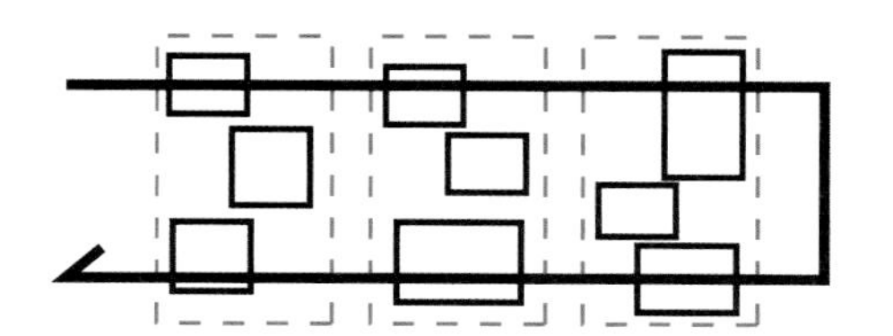

传统岭南园林清晖园内部也分为多种空间，人们在其中游览时会自发地形成一条流线，这条流线贯穿建筑始终，使人在游览时尽量不走重复路线

图3-49 “折叠与起伏”步行路线组织原则

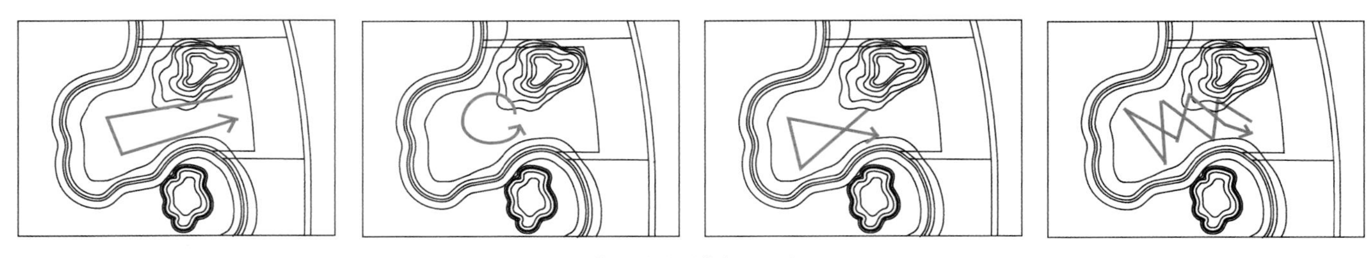

“折叠与起伏”布局原则

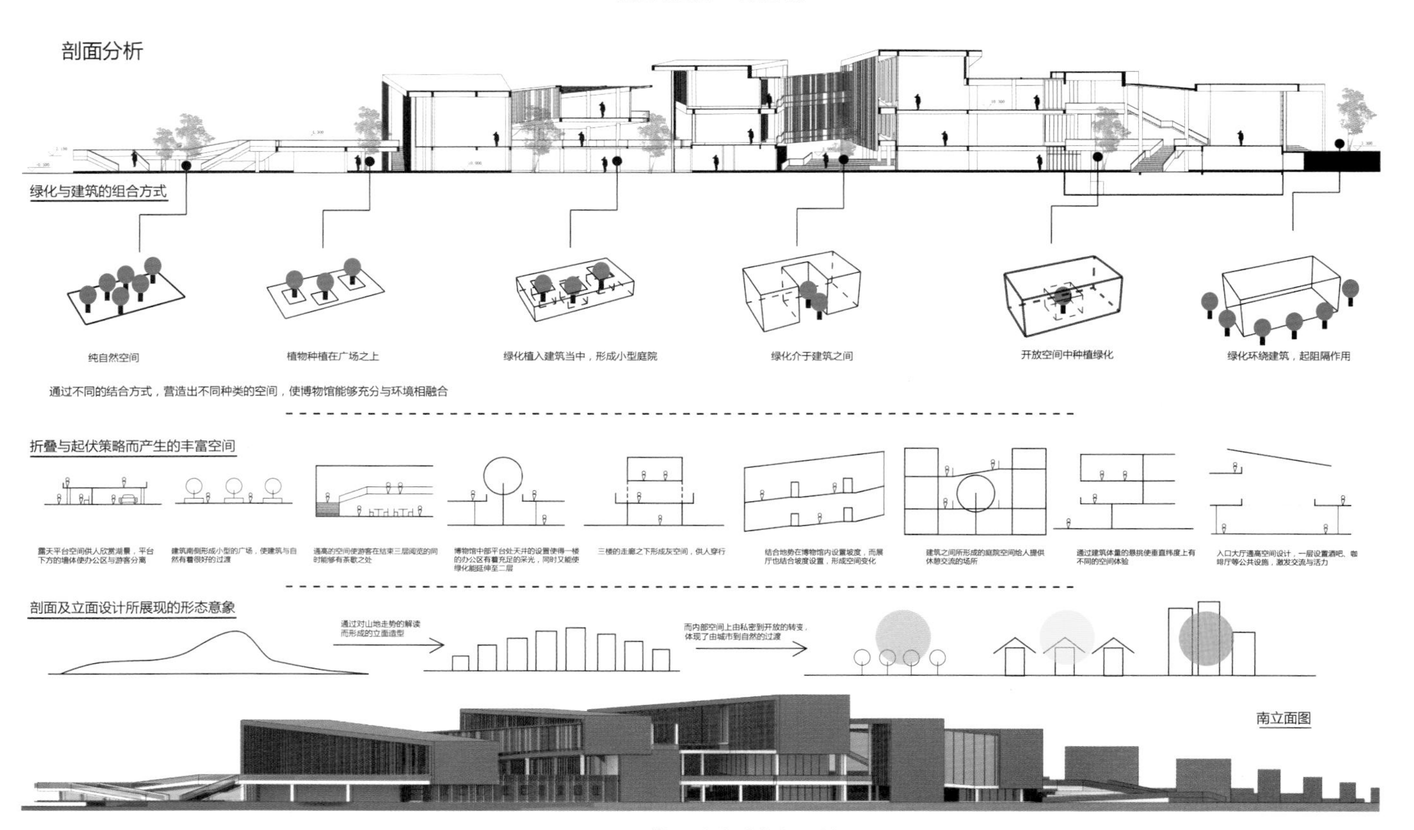

图3-50　“折叠与起伏”空间分析

图3-51 “折叠与起伏”空间序列

3.3.4 “生态结”：高校图书馆设计

设计场地位于高校的主轴线上，地处校园的核心地带，景观资源丰富，地形较复杂；设计师因而认为，图书馆作为大学重要的学习场所，应与各种功能区域建立多向的联系，并成为一个融合交通、景观、学习及交往多种功能的场所，以提升校园的品质和活力。

基于此理念，设计结合地形，在不同标高形成边界模糊、形态有机的建筑空间，容纳来自各个方向的人流活动，并使建筑有效地融入自然环境中。与此同时，设计引入若干庭院，为建筑内部带来自然光线的同时，也进一步使建筑空间自然化，以创造一个绿色生态的学习环境，使学生寓学习于自然景观中，发挥自然的生态教育与净化心灵的作用。藉着自然化的设计手段，图书馆也在周边不同设施之间建立内在的联系，从而成为校园中激发活力与互动的交汇点，为传统的高校图书馆设计带来观念上的革新（图3-52~图3-57）。

3.3.4 Eco-node: A University Library Design

Being part of the university axis, the site is located at the core of the campus. Since the landscape resources are rich and the topography is relatively complex, the designer believed that the library as an important learning place, should establish a multi-directional connection with other facilities, and should bear various functions, including circulation, learning, interaction and access to nature to enhance the quality and vitality of the campus.

Based on this concept, the designer, by taking into account the variation of terrain, created a library with an organic shape and blurred boundaries at different levels, accommodating circulation from different directions, as well as merging itself effectively with the natural environment. Meanwhile, through introducing a series of courtyards into the library, natural light is allowed to project into the building, and thus an ecological learning environment is created. Inside this environment, students are able to study by integrating with natural landscape, and consequently nature plays an important role in education and spiritual purification.

Deemed an eco-node inside the campus, the library has established a coherent connection with the surrounding facilities, and thus can become a vivid place, stimulating activities and interaction among students. Through employing a number of N-architecture design strategies, this library has brought innovation to the traditional design concept in some ways.

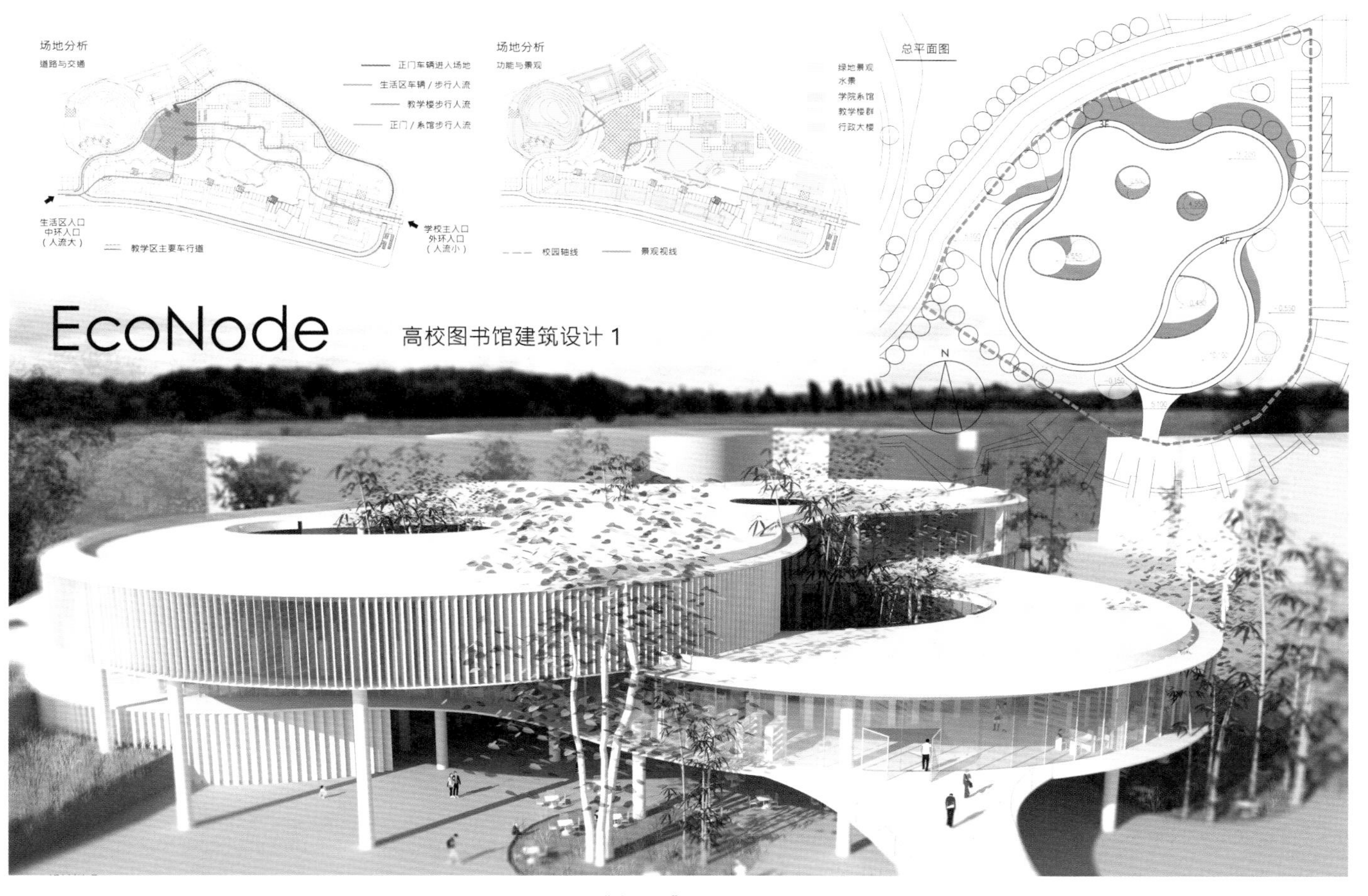

图3-52 “生态结”场地分析及生成概念

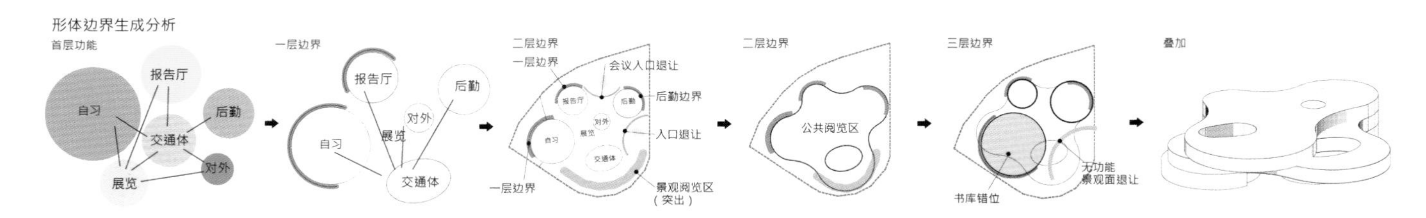

图3-53 “生态结”的设计模型

概念分析

各功能组团与场地的边界信息

正门/系馆

主要交通环路

教学楼

可能的行走路线

场地特征

- 多向性/无方向性
- 植被丰富、绿化优良
- 处于校园中心位置

图书馆概念

- 多向性/无方向性
- 尽可能开发外环境绿化与内庭院植被

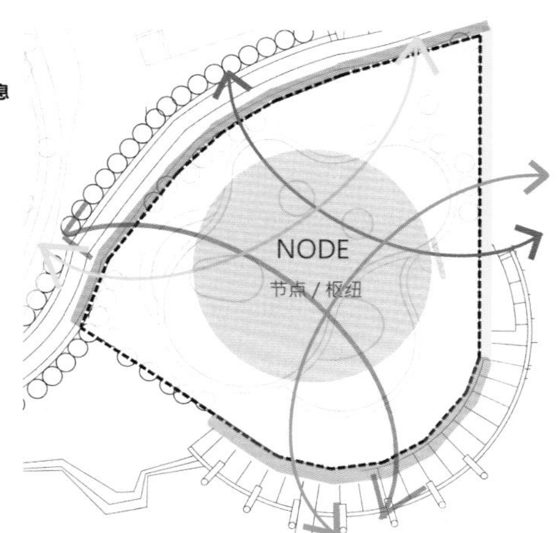

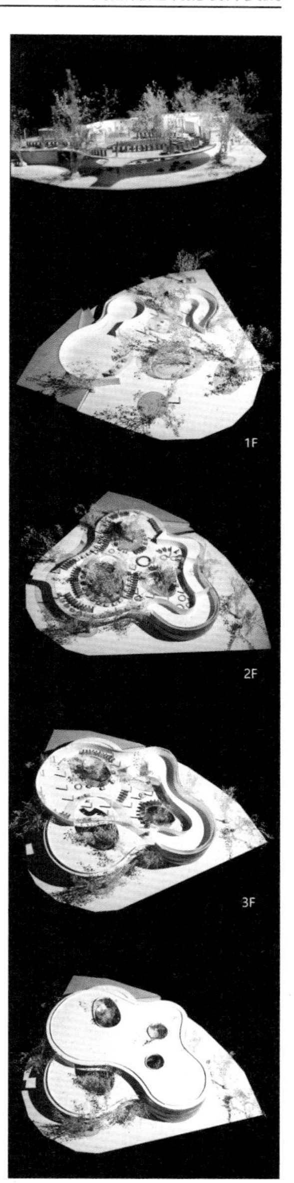

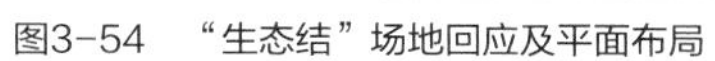

图3-54　“生态结”场地回应及平面布局

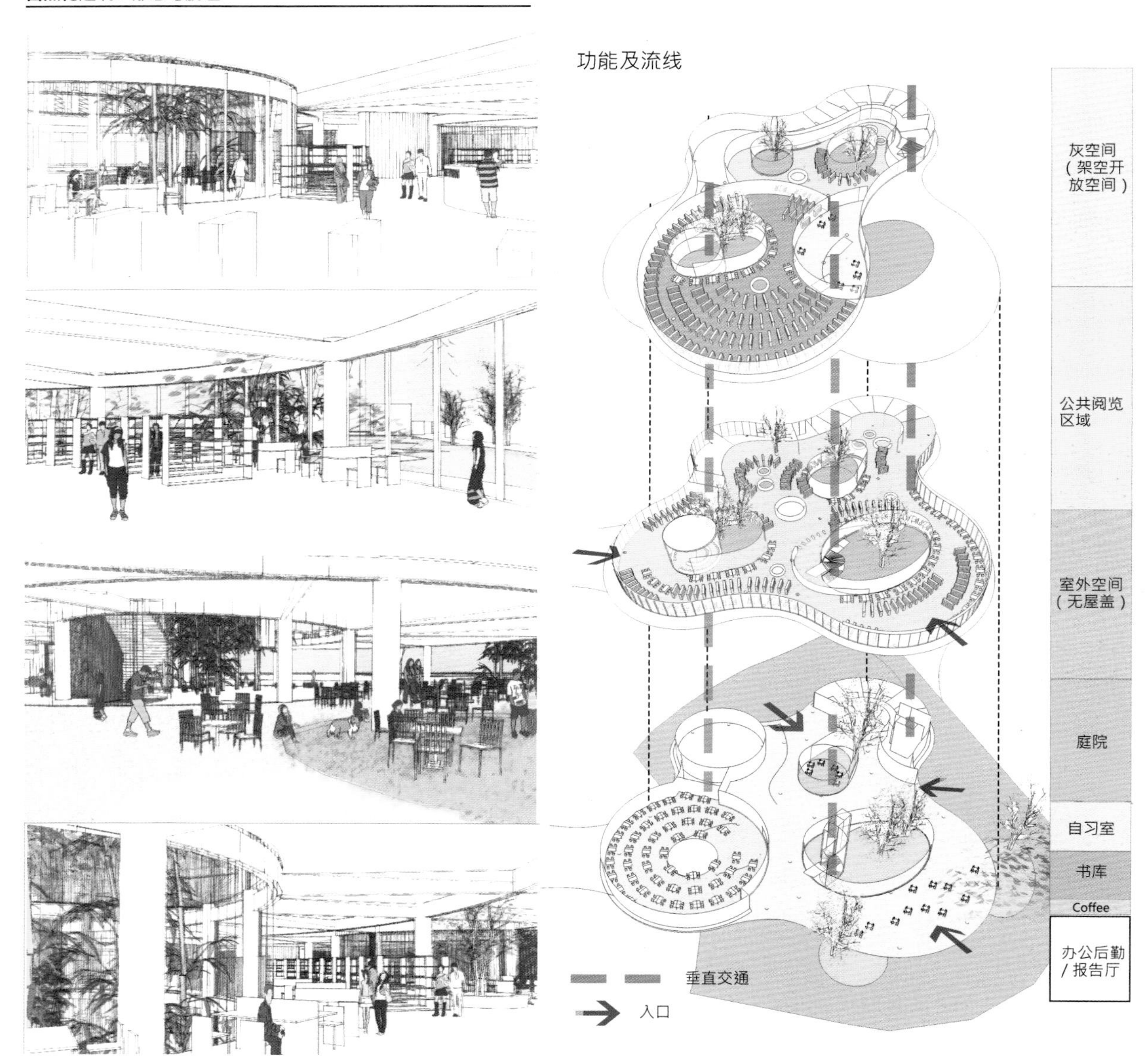

图3-55 “生态结”功能布局及流线

图3-56 “生态结”的设计模型

图3-57 “生态结”剖面与立面图

3.4 功能重构

当功能问题成为建筑设计实践和建筑教育中的主要内容被反复强调时，建筑学所面临的挑战就是如何能超越基于知识的悖论，探索功能表象背后更深层次的逻辑和理论依据。因此，对建筑功能的含义进行重新思考是有必要的。当我们尝试从不同的角度去看待功能问题时，设计将会产生更多的可能性。

以偏向静止的角度来看，功能是个语汇，是空间的名称和要求；但从动态的角度，它则隐含着特定的空间行为模式与社会关系。功能是人们通过长期实践总结出来的社会关系的反映；因此，当构建新的社会关系成为一种需要时，对功能进行重构就成为一种设计手段。而这种设计模式，有利于设计从一种基于知识的思维转化为基于研究和知性的行为。在功能重构的过程中，自然也被视作功能元件，在整个空间系统中扮演构建社会关系的角色。与生态社会策略有所不同，这种空间类型的产生是主要从功能再生开始的。

3.4 Reprogramming

When functionalism is highlighted in both architectural education and practice, a challenge that we face is going beyond the knowledge-based paradox and exploring a deeper logic or rationale behind the functional representations. Therefore, to rethink the meaning of architectural functions is necessary. When looking at the functional issues from different perspectives, architectural design practice is able to produce more possibilities.

From a relatively static perspective, function is a word, the name or the requirement of a specific space; but from a dynamic point of view, it implies specific spatial behavioural patterns and certain social relationships. In truth, programming is a reflection of social order and relationships that people have drawn up through their long-term practice. When there is a need to reconfigure the social relationship inside a building, reprogramming can be a design tool and a way of creation. Through reprogramming, design practice can be transformed from a knowledge-based way of thinking into a more research-based and intellectual-based type of behaviour. During the process of reprogramming, natural landscape is always regarded as a functional element, playing an important role in building up the social logic of a spatial system. Unlike eco-society design ideas, this type of design approach begins with functional regeneration.

3.4.1 “空间频谱”：高校图书馆设计

这里所展示的是一个高校图书馆设计案例。设计首先对大学生的学习行为进行再思考，并探讨不同学习模式与空间使用之间的相互关系。设计师认为，空间对使用者的行为有影响作用，但使用者也可以通过特定的行为模式对空间性质进行界定。因应空间里发生的行为活动，空间因而产生不同的活跃度与透明度。

基于这种理念，设计者对图书馆的功能进行重构，发展出一套功能频谱，并以一立体交通系统把它们联系起来；例如，阅读空间是可达性最强、活跃度最高的空间，而研究空间次之，等等。阅读空间因而被设计成悬浮在空间中的体量，形成建筑的视觉焦点；而自然空间如庭院穿插其间，成为学习空间的一部分，从而实现了建筑自然化与自然建筑化的设计目的（图 3-58~ 图 3-63）。

3.4.1 Spatial Spectrum: A University Library Design

The example presented here is a university library design. Through analyzing the learning behaviour of students, the designer put his focus on the impact of different learning patterns on the spatial uses and further drew the conclusion that the interaction between people and space is dynamic. Spatial quality can influence users' behaviour, but occupants' behavioural patterns can also define the character of a space, further leading to different degrees of its activity and transparency.

In this sense, the programme of the library was reconstructed and developed into a set of spectrums, linked by a three-dimensional circulation system; for example, the reading rooms are ranked as the most accessible and vivid space, followed by research space, and so on. The reading space is then designed to be floating inside the volume, forming a focal point. The natural setting, like courtyards, are integrated with reading and research spaces, helping achieve harmony between architecture and nature.

图3-58 “空间频谱”设计模型

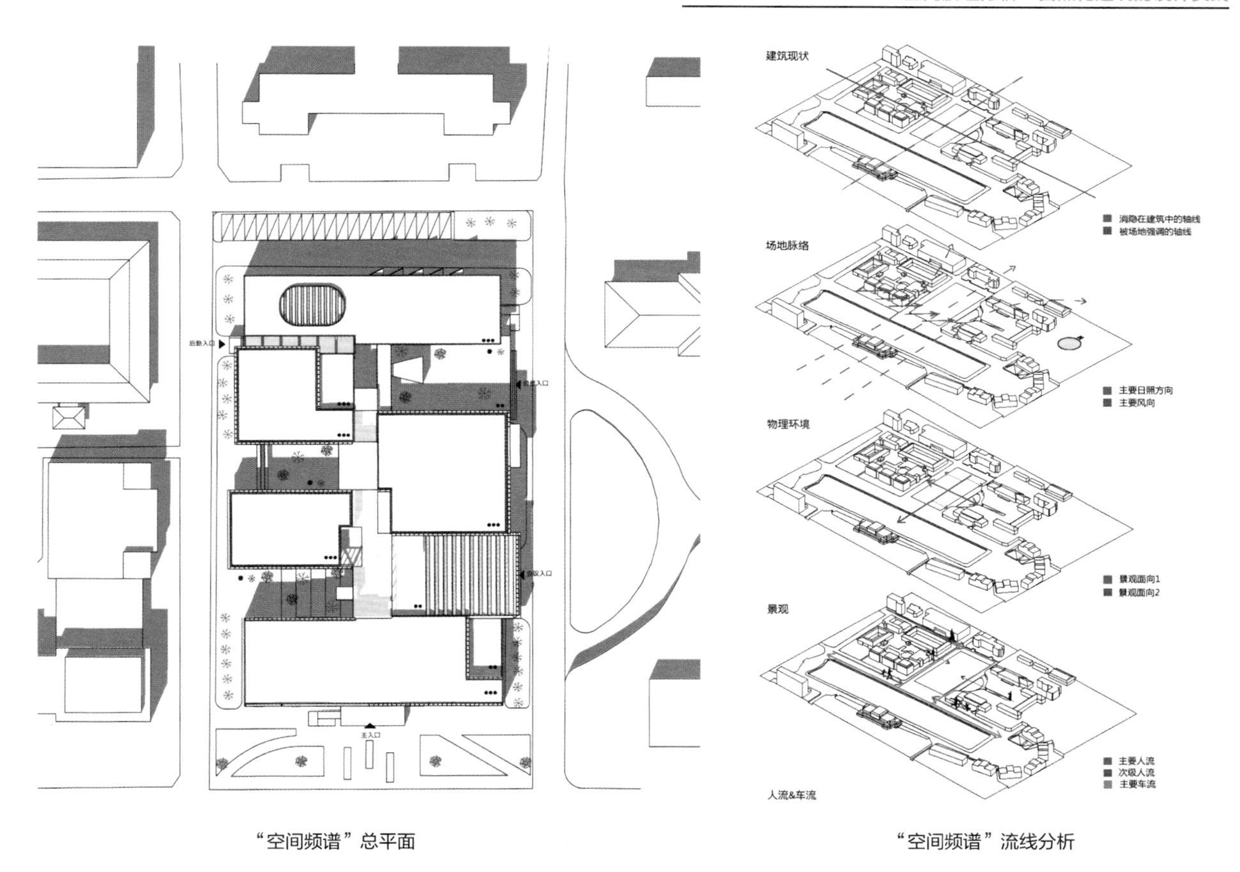

“空间频谱”总平面

“空间频谱”流线分析

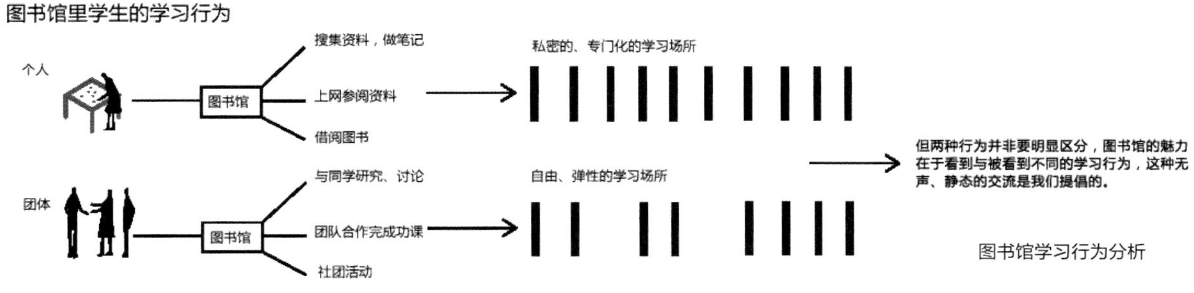

图书馆学习行为分析

图3-59　“空间频谱”设计概念

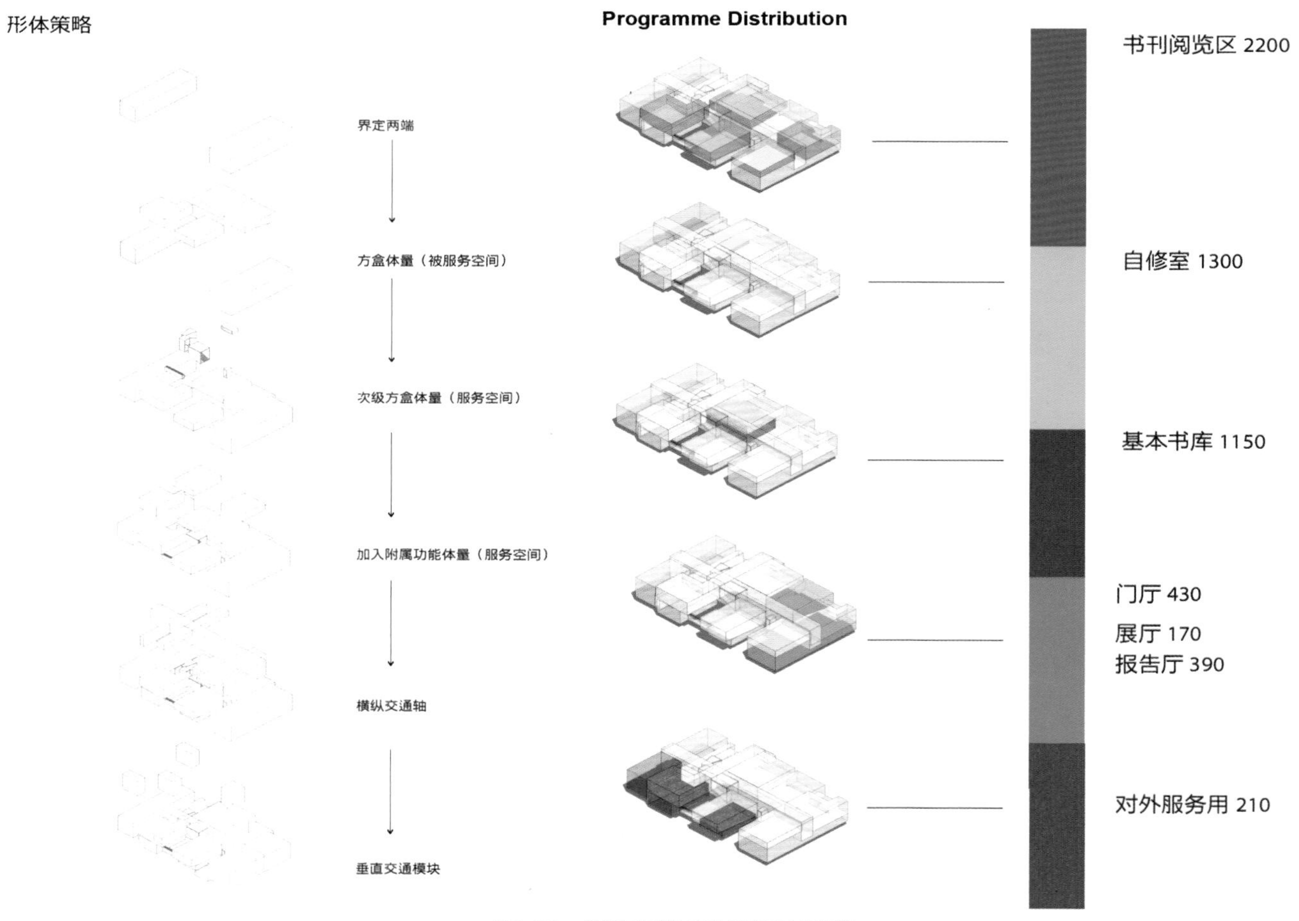

图3-60 “空间频谱”形体策略及功能频谱

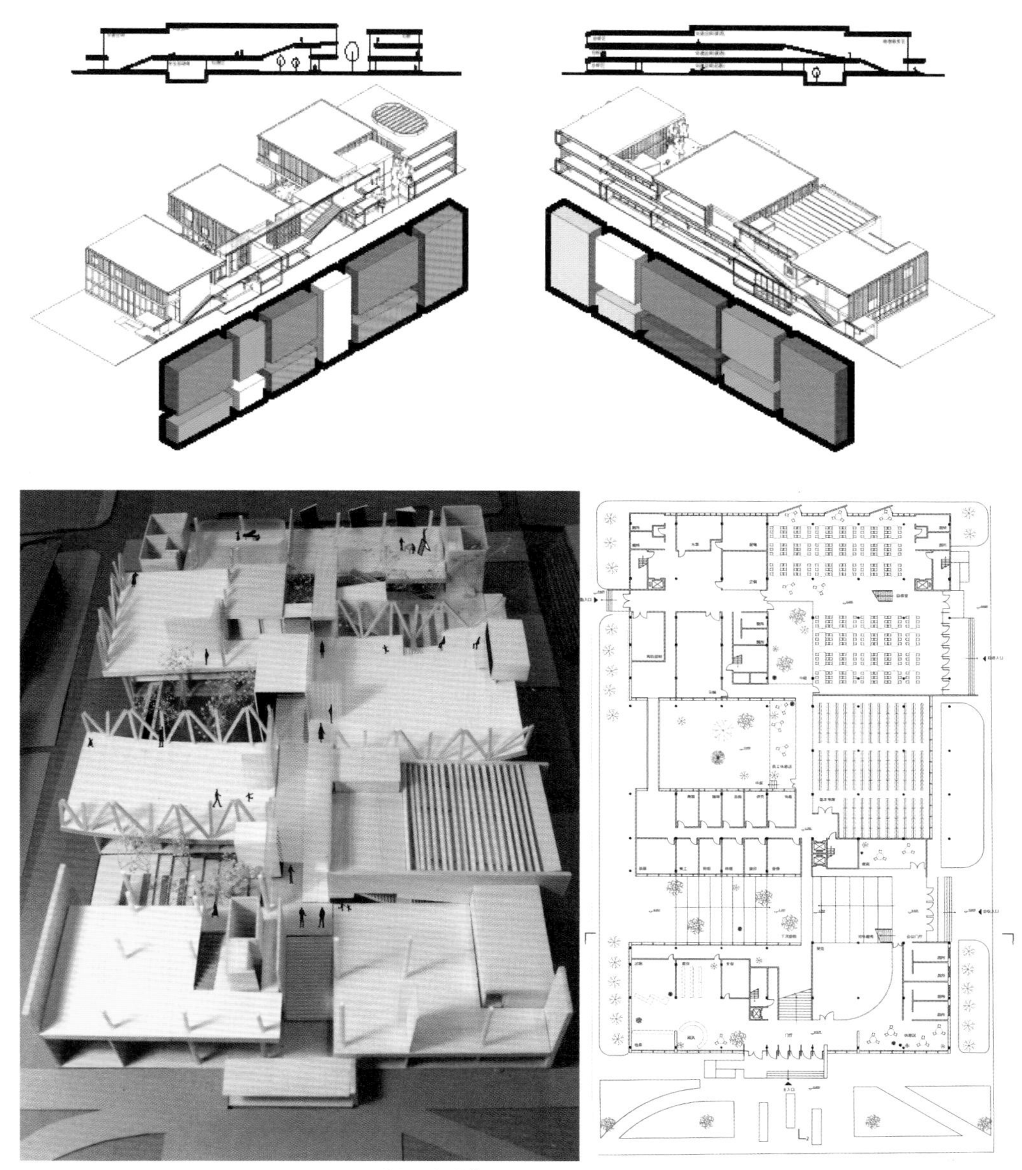

图3-61 “空间频谱”首层平面图

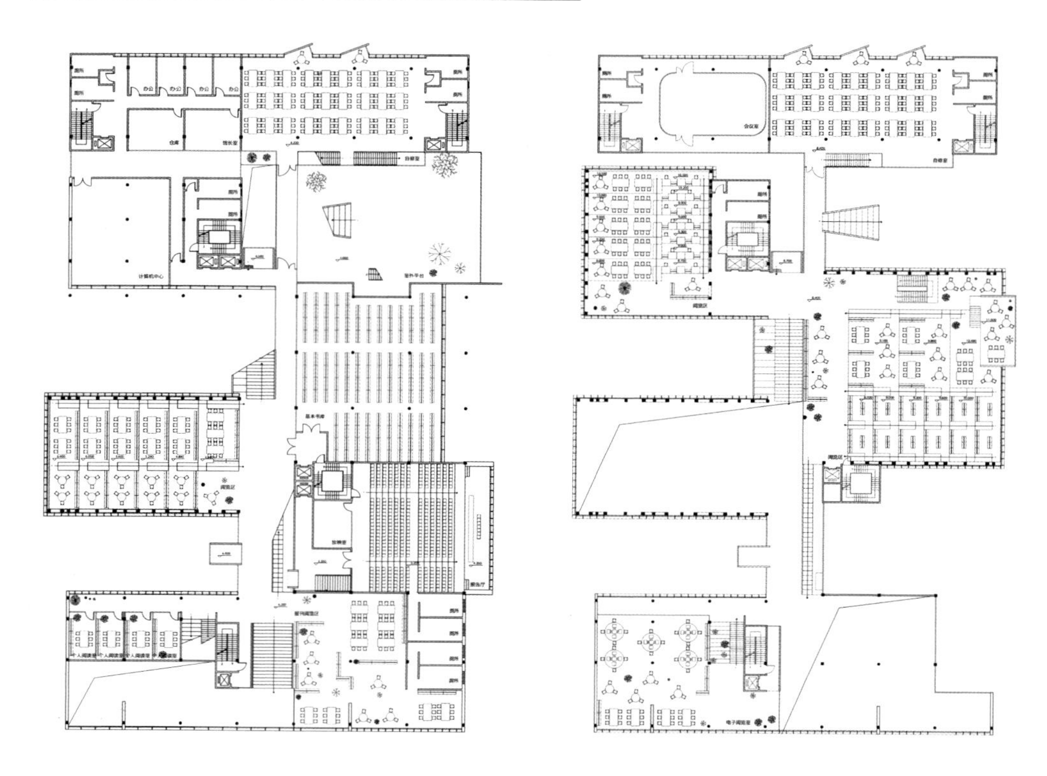

图3-62　“空间频谱”二、三层平面图

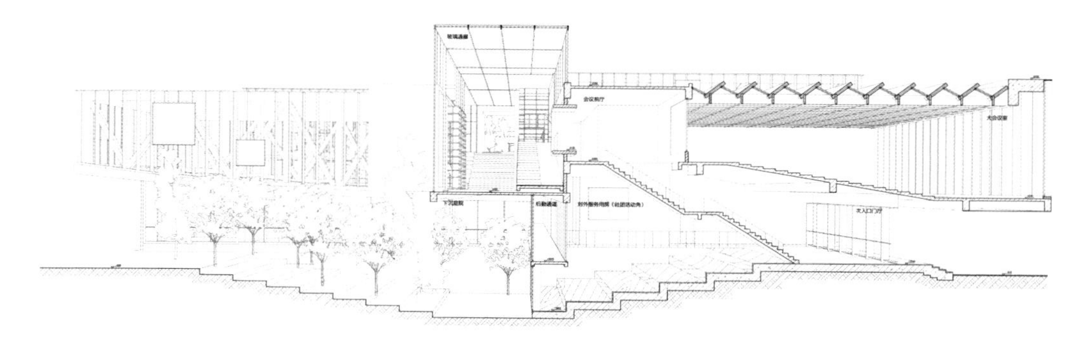

图3-63　“空间频谱”剖面图

3.5 小结

本章的作品主要展示了几种具有代表性的自然化建筑设计思路。得益于自然界的丰富及深邃，自然对建筑实践的启示将是无穷无尽的。藉着人类不懈的探索与努力，相信自然化建筑将进一步发展，未来将会有更丰硕的设计成果展现在我们面前。

3.5 Epilogue

The main works introduced by this chapter only illustrated several representative design ideas of N-architecture. Owing to the richness and depth of nature, the enlightenment of nature in architectural practice will be endless. By means of relentless exploration of designers, it is fair to predict that N-architecture will be further developed, with more fruitful design works emerging in architectural fields.

参考文献 Bibliography

[1] ALEXANDER C. A City is Not a Tree [J]. Architectural Forum, 1965, 122(1): 58-62.

[2] STAN A. Points + Lines: Diagrams and Projects for the City [M]. New York: Princeton Architectural Press, 1999.

[3] STAN A, MCQUADE M. Landform Building: Architecture's New Terrain [M]. Baden, Switzerland: Lars Müller Publishers; Princeton, N.J.: Princeton University School of Architecture, c2011.

[4] BENJAMIN A. Reiser + Umemoto: Recent Projects [M]. London: Academy Editions, 1998.

[5] Complexity and Consistency: Foreign Office Architects (FOA) 1996-2003 [J]. EL Croquis, 2003 (115/116).

[6] COONES P. One Landscape or Many? A Geographical Perspective [J]. Landscape History, 1985.

[7] CORBUSIER L. Towards A New Architecture [M]. New York: Dover, 1931, 1986.

[8] EDMUND N B. Design of Cities [M]. London: Thames & Hudson, 1974.

[9] EISENMAN P. Diagrams of Anteriority [M]. Diagram Dairies, Universe, 1999.

[10] AMBASZ E, RILEY T. Emilio Ambasz: Architectura e Design [M]. Introductione di Terence Riley. Milano: Electa, c1993.

[11] Enric Miralles: Benedetta Tagliabue 2000-2009 [J]. EL Croquis, 2009(144).

[12] FABRIZI M. The Building is the City: Le Corbusier's Unbuilt Hospital [J]. In Socks Studio. http://socks-studio.com/2014/05/18/the-building-is-the-city-le-corbusiers-unbuilt-hospital-in-venice/ [2016-10-10].

[13] FALCONER K J. Fractal Geometry: Mathematical Foundations and Applications [M]. Hoboken: John Wiley & Sons Inc., 2014.

[14] FRAMPTON K (edit.). Tadao Ando: Buildings, Projects, Writings [M]. New York, N.Y.: Rizzoli, 1984.

[15] FRANCIS D K C. Architecture: Form, Space and Order [M]. New York: Van Nostrand Reinhold, 1979.

[16] HART S. Ecoarchitecture: the Work of Ken Yeang [M]. Hoboken, N.J.: Wiley, 2011.

[17] HENSEL M U, TURKO J P. Grounds and Envelopes: Reshaping Architecture and the Built Environment [M]. Abingdon, Oxon; New York: Routledge, 2015.

[18] JACKSON B. Discovering the Vernacular Landscape [M]. New Haven, MA: Yale University Press, 1984.

[19] JENCKS C. The Architecture of the Jumping Universe: A Polemic: How Complexity Science is Changing Architecture and Culture [M]. London: Academy Editions, 1997.

[20] JODIDIO P. Ando: Complete Works [M]. London: Taschen, 2012.

[21] KANT I. Critique of Judgment [M]. Translated by Werner S. Pluhar. Indianapolis: Hackett Publishing Company, 1987.

[22] KOOLHAAS R, MAU B, SIGLER J (edit.). Small, Medium, Large, Extra-large: Office for Metropolitan Architecture [M]. New York, N.Y.: Monacelli Press, 1998.

[23] MANDELBROT B. B. The Fractal Geometry of Nature [M]. San Francisco: Freeman, 1982.

[24] MCHARG I. L. Design with Nature [M]. New York: Natural History Press, 1969.

[25] MCHARG I. L. To Heal the Earth: Selected Writings of Ian L. McHarg [M]. Frederick R. Steiner, Washington D.C. & Covelo, CA: Island Press, 1998.

[26] NAVEH Z, LIEBERMAN A. S. Landscape Ecology: Theory and Application [M]. New York: Springer Science + Business Media, LLC, 1984.

[27] NONELL J B. Antonio Gaudi: Master Architect [M]. New York: Abbeville Press. 2004.

[28] OMA/Rem Koolhaas 1987-1998 [J]. EL Croquis, 2002.

[29] PEARSON D. The Breaking Wave: New Organic Architecture [M]. Stroud: Gaia, 2001.

[30] PREIFFER B B. Frank Lloyd Wright, 1867-1959: Building for Democracy [M]. Los Angeles: Taschen, 2004.

[31] ROCCA A. Natural Architecture [M]. New York: Princeton Architectural Press; Milan, Italy: 22 Pub., c2007.

[32] TSCHUMI B. Event-cities: Praxis [M]. Cambridge, Mass.: MIT Press, c1994.

[33] TSCHUMI B. Architecture and Disjunction [M]. Cambridge, Mass.: MIT Press, c1994.

[34] WRIGHT F L. The Natural House [M]. London: Pitman, 1971.

[35] YEANG K. Tropical urban Regionalism: Building in a South-East Asian City [M]. Singapore: Concept Media, 1987.

[36] YU K J,PADUA M (edit.). The Art of Survival: Recovering Landscape Architecture [M]. Mulgrave, Victoria: Images Publishing Group, c2006.

[37] Zaha Hadid 1996-2001 [J]. EL Croquis, 2001(103): 44-51.

[38] 俞孔坚 . 景观：文化、生态与感知 [M]. 北京：科学出版社 ,1998.

[39] 俞孔坚 . 生物与文化基于上的图式——风水与理想景观的深层意义 [M]. 台湾：田园文化出版社 , 1998.

[40] 俞孔坚 . 景观的含义 [J]. 时代建筑，2002(1)：14–17.

图形索引与注释　Index & Credits of Graphs

图 1-1　https://www.pinterest.com/pin/135741376244583838/ [2016-10-10]

图 1-2　http://d1dzqwexhp5ztx.cloudfront.net/imageRepo/4/0/81/262/819/Central_Park_Arial_View_S.jpg. [2016-10-10]

图 1-3　http://www.botanichka.ru/wp-content/uploads/2016/12/english-style-garden-2.jpg [2016-10-10]

图 1-4　https://architizer.com/projects/philadelphia-navy-yards-central-green/media/1416540/ [2016-10-10]

图 1-5　http://kienviet.net/wp-content/uploads/2015/09/Jeju-Hill-Hotel-Resort-1.jpg [2017-06-01]

图1-6　Giambattista Nolli. Map of Rome, 1748-Earth Sciences & Map Library-University of California,Berkeley.

http://www.lib.berkeley.edu/EART/maps/nolli_06.jpg [2016-10-10]

图 1-7　Le Corbusier's 'Green space' plan for St. Die http://4.bp.blogspot.com/yhcMvVjsb24/UBbf39elnZI/AAAAAAAAAJk/nWYDUtI0i40/s1600/Urbanism-2_1_Collage+City+FIgure+ground.jpg [2016-10-10]

图 1-8　https://s3.amazonaws.com/classconnection/680/flashcards/11483680/jpg/crystal_palace_lodon132907925995 81364586295968-15491BA742741FB77A2.jpg [2017-05-30]

https://upload.wikimedia.org/wikipedia/commons/e/e1/Bauhaus.JPG/

https://upload.wikimedia.org/wikipedia/commons/b/bb/Bauhaus_Dessau-001.jpg [2017-05-30]

https://userscontent2.emaze.com/images/efd5f353-1e7a-443e-afe7-e7e1847ce780/f6fcab6-a89b-4626-9841-64d446abb9dd.jpg [2017-05-30]

图 1-9　http://socks-studio.com/2014/05/18/the-building-is-the-city-le-corbusiers-unbuilt-hospital-in-venice/ [2017-05-30]

http://predmet.fa.uni-lj.si/siwinds/s1/u1/su6/img/s1_u1_su6_p2_1.jpg

图 1-10　http://www.europan-europe.eu/media/default/0001/13/39b6c310e1822d23752a7eec2888d1e33e2e8421.png[2017-05-30]

图 1-11　ALLEN S. Points + Lines: Diagrams and Projects for the City [M]. New York: Princeton Architectural Press, 1999: 98.

图 1-12　ALLEN S. Points + Lines: Diagrams and Projects for the City [M]. New York: Princeton Architectural Press, 1999: 99.

https://cdn1.pri.org/sites/default/files/story/images/RTR1KSFP_0.jpg [2017-05-30]

图 1-13　JENSKS C. The Architecture of the Jumping Universe-A Polemic: How Complexity Science is Changing Architecture and Culture [M]. London: Academy Editions, 1997: 30.

图1-14 FALCONER K. Fractal Geometry: Mathematical Foundations and Applications [M]. Hoboken: John Wiley & Sons Inc., 2014.

图2-1 BENJAMIN A. Reiser + Umemoto: Recent Projects [M]. London: Academy Editions, 1998.

图2-2 https://cerijmdavies.files.wordpress.com/2013/10/gaudi1.jpg [2017-05-30]

http://free4kwallpaper.com/wp-content/uploads/2016/02/Stunning-Barcelona-Spain-4K-Wallpapers.jpg [2017-05-30]

https://finnternational.files.wordpress.com/2012/11/2012-09-07-11-45-19.jpg [2017-05-30]

图2-3 http://thefunambulistdotnet.files.wordpress.com/2011/08/enric-miralles-the-funambulist-5.jpg [2017-05-30]

http://1.bp.blogspot.com/-8DTdY-yGEUw/T5iZtF757MI/AAAAAAAAAfs/wp_i5-CgKg8/s1600/Screen+shot+2012-04-25+at+8.27.46+PM.png [2017-05-30]

http://www.landezine.com/wp-content/uploads/2009/06/051.jpg [2017-05-30]

图2-4(a,b) JODIDIO P. Ando: Complete Works [M]. Hong Kong: Taschen, 2007.

http://www.archilibra.com/thesis/case_studies/habitat/section.jpg [2017-05-30]

图2-5(a,b) Emilio Ambasz: Architectura e Design [M]. Introductione di Terence Riley. Milano: Electa, c1993: 80-81.

图2-6 http://www.wright-house.com/frank-lloyd-wright/fallingwater-pictures/large-fallingwater-photos/high-resolution/falling-water-fall-house-L.jpg [2017-05-30]

http://www.wright-house.com/frank-lloyd-wright/fallingwater-pictures/large-fallingwater-photos/high-resolution/7x-front-door-east-end-L.jpg [2017-05-30]

http://all-that-is-interesting.com/wordpress/wp-content/uploads/2012/11/MostFamousDesignsof Frank WrightFallingwater12.jpg [2017-05-30]

图2-7 Zaha Hadid 1996-2001[J]. EL Croquis, 2001(103): 44-51.

图2-8 http://oma.eu/projects/jussieu-two-libraries [2017-05-30]

图2-9 http://images.adsttc.com/media/images/5441/0b37/c07a/801f/e700/049a/large_jpg/5420792ec07a800de500000e_ad-classics-yokohama-international-passenger-terminal-foreign-office-architects-foa-_yipt-0802-satoru_mishima-05.jpg?1413548844 [2017-05-30]

http://proyectoblogspace.com/wp-content/uploads/2011/10/projekt-foa-yokohama-06.jpg [2017-05-30] https://stephenvaradyarchitraveller.files.wordpress.com/2016/04/yokohama-port-terminal-by-foreign-of fice-architects-69_stephen-varady-photo-c2a9.jpg [2017-05-30]

http://a1.images.divisare.com/images/f_auto,q_auto/v1459428535/hehj6tz8ughpslfeet0s/foa-azpml-farshid-moussavi-architecture-valerie-bennett-yokohama-international-port-terminal.jpg [2017-05-30]

图 2-10　http://assets.inhabitat.com/wp-content/blogs.dir/1/files/2015/07/Crawick-Multiverse-Scotland-4.jpg [2017-05-30]

图 2-11　https://www.architecturalpapers.ch/images/articles/9_1_w1000h600.jpg [2017-05-30]

图 2-12　http://cche.ch/wp-content/uploads/2016/10/AP1-1140x640.jpg [2017-05-30]

图 2-13、图 2-14　作者设计及绘制，合作者：Nacho Carbo、Aimon Tabony、Mark Mat

图 2-15　HART S. Ecoarchitecture: the Work of Ken Yeang [M]. Hoboken, N.J.: Wiley, 2011: 192-199.

图 3-1　作者自绘

图 3-2 ~图 3-6　作者设计及绘制，合作者：Mak Fai

图 3-7 ~图 3-11　作者任设计指导，设计及绘制：张雯

图 3-12 ~图 3-16　作者设计及绘制

图 3-17　作者自绘

图 3-18 ~图 3-22　作者任设计指导，设计及绘制：杨森博、林丹薇

图 3-23 ~图 3-31　作者任设计指导，设计及绘制：陈一叶、李桐

图 3-32 ~图 3-36　作者任设计指导，设计及绘制：冯思蕴、鲍捷

图 3-37 ~图 3-42　作者设计及绘制，合作者：Ducan Chan、David Chan 等

图 3-43 ~图 3-45　作者任设计指导，设计及绘制：万斯涵

图 3-46 ~图 3-51　作者任设计指导，设计及绘制：仇普钊、陈曦

图 3-52 ~图 3-57　作者任设计指导，设计及绘制：沈文婕

图 3-58 ~图 3-63　作者任设计指导，设计及绘制：邬皓南

图书在版编目（CIP）数据

自然化建筑：形态与肌理：汉英对照/凌晓红著. —广州：华南理工大学出版社，2017.6
ISBN 978-7-5623-5281-5

Ⅰ. ①自… Ⅱ. ①凌… Ⅲ. ①建筑设计-汉、英 Ⅳ. ①TU2

中国版本图书馆CIP数据核字(2017)第104101号

自然化建筑：形态与肌理
凌晓红 著

出 版 人：卢家明
出版发行：华南理工大学出版社
（广州五山华南理工大学17号楼，邮编510640）
http://www.scutpress.com.cn E-mail: scutc13@scut.edu.cn
营销部电话：020-87113487 87111048（传真）
责任编辑：刘志秋 赖淑华
印 刷 者：广州市骏迪印务有限公司
开 本：787mm×1092mm 1/12 **印张：**12 **字数：**222千
版 次：2017年6月第1版 2017年6月第1次印刷
定 价：128.00元